THE SOCIOBIOLOGY OF SEXUAL AND REPRODUCTIVE STRATEGIES

THE SOCIOBIOLOGY OF SEXUAL AND REPRODUCTIVE STRATEGIES

Edited by

ANNE E. RASA
Department of Zoology,
University of Pretoria

CHRISTIAN VOGEL
and
ECKART VOLAND
Institute for Anthropology,
University of Göttingen

LONDON
NEW YORK
Chapman and Hall

First published in 1989 by
Chapman and Hall Ltd
11 New Fetter Lane, London EC4P 4EE
Published in the USA by
Chapman and Hall
29 West 35th Street, New York NY 10001

Typeset in 10/12pt Plantin Light by
Leaper & Gard Ltd, Bristol
Printed in Great Britain at the
University Press, Cambridge

ISBN 0 412 33780 0

British Library Cataloguing in Publication Data
The Sociobiology of sexual and reproductive strategies
  1. Organisms. Reproduction
I. Rasa, Anne, 1940– II. Vogel, Christian, 1933
III. Voland, Eckart, 1949–
574.1′6

ISBN 0 412 33780 0

Library of Congress Cataloging-in-Publication Data
The Sociobiology of sexual and reproductive strategies/edited by
  Anne E. Rasa, Christian Vogel, and Eckart Voland.
     p.   cm.
    Based in part on a meeting held Jan. 1986 in Gottingen
(FRG) and sponsored by the European Sociobiological
Society.
    Includes bibliographies and index.
    ISBN 0-412-33780-0
    1. Sexual behaviour in animals–Congresses.   2. Sex.
3. Social behavior in animals–Congresses.
4. Interpersonal relations–Congresses.   5. Sociobiology–
Congresses. 6. Reproduction–Congresses.   I. Rasa,
Anne, 1940– .   II. Vogel, Christian, 1933–   .   III.
Voland, Eckart, 1949–   .   IV. European Sociobiological
Society.
    [DNLM: 1. Reproduction–congresses.   2. Sex
Behavior–congresses.   3. Sex Behavior, Animal–
congresses.   4. Social Behavior–congresses.   HQ 21
S6795 1986]
QL761.S6   1989
591.56–dc19
DNLM/DLC
for Library of Congress   88-38328
                                             CIP

Contents

PART ONE

Eco-ethological aspects of sexuality and reproduction in animal populations

Contents

Contributors

Harris Bernstein
Department of Microbiology and Immunology, University of Arizona, Tucson, USA

Nicholas G. Blurton
Graduate School of Education and Department of Anthropology and Psychiatry, University of California, Los Angeles, USA

Robin I.M. Dunbar
Department of Anthropology, University College London, London, UK

Bruce Ellis
San Francisco, USA

Claudia Engel
Institut für Anthropologie, Universität Göttingen, Göttingen, FRG

Mark V. Flinn
Department of Anthropology, University of Missouri, Columbia, USA

Karl Grammer
Forschungsstelle für Humanethologie, Max-Planck-Institut für Verhaltensphysiologie, Seewiesen, FRG

Frederic A. Hopf
Optical Science Center, University of Arizona, Tucson, USA

Colin Irwin
Department of Sociology and Social Anthropology, Dalhousie University, Halifax, Canada

Jutta Kuester
Affenberg Salem, Salem, FRG

Jürg Lamprecht
Max-Planck-Institut für Verhaltensphysiologie, Seewiesen, FRG

Contributors

K. Eduard Linsenmair — Zoologisches Institut, Universität Würzburg, Röntgenring, Würzburg, FRG

Richard E. Michod — Department of Ecology and Evolutionary Biology, University of Arizona, Tucson, USA

Andreas Paul — Affenberg Salem, Salem, FRG

Anne E. Rasa — Department of Zoology, Faculty of Mathematics and Science, University of Pretoria, Republic of South Africa

Wulf Schiefenhövel — Forschungsstelle für Humanethologie, Max-Planck-Institut für Verhaltensphysiologie, Seewiesen, FRG

Volker Sommer — Institut für Anthropologie, Universität Göttingen, Göttingen, FRG

Donald Symons — Department of Anthropology, University of California, Santa Barbara, USA

Christian Vogel — Institut für Anthropologie, Universität Göttingen, Göttingen, FRG

Eckart Voland — Institut für Anthropologie, Universität Göttingen, Göttingen, FRG

Preface

In January 1986 the European Sociobiological Society (ESS) held a scientific meeting in Göttingen (FRG) on sexual and reproductive strategies in animals and humans. Its purpose was to bring together people working on as many different aspects and as many different organisms as possible, in order to stimulate discussion and exchange of information on these topics. During a three day conference a rich programme of papers was presented by scientists of various disciplines from all over the world. The spectrum of these presentations ranged from the origins of sexual reproduction at the level of the gene up to the highly sophisticated reproductive strategies of human cultures. Papers varied from the presentation of empirical data on sexual and reproductive strategies in organisms as different as isopods and humans, to the proposal of new theoretical concepts regarding reproduction itself. This wide variety of topics looked at from genetical, ethological, sociobiological, and anthropological perspectives generated stimulating discussions which were particularly effective in uncovering the similarities as well as the differences between the strategies of reproductive competition and propagation of genetic material both by animals and man.

Since the meeting appeared to have been creative the participants voted to publish a book of selected papers. The editors decided to request some supplementary contributions from authors prominent in the field in order to enhance public interest.

We would like to convey our thanks to all participants for the efficient running of the conference and their stimulation of an atmosphere conducive to an innovative communication among scientists of different disciplines. We would also like to thank the members of the ESS board for their excellent support of the Göttingen meeting and their collaborative aid for the publication of this volume.

Anne E. Rasa, Christian Vogel, Eckart Voland
Göttingen, September 1988

ix

Introduction: on sexual and reproductive strategies

Christian Vogel

All organisms are shaped by natural selection. Since natural selection operates through differential reproduction, this makes reproduction the key phenomenon of evolution. Hence, all organisms compete for their own reproductive success which is, in general, the most effective means of maximizing personal fitness. However, under certain conditions some individuals of socially living animals may postpone or even forego their own reproduction in order to maximize their inclusive fitness, for instance, by taking the role of 'helpers at the nest', i.e. helping closely related individuals to raise their offspring successfully. Thus, we may find highly sophisticated strategies of transferring as many replicators of 'own' genes to the next generation as possible. Of course, animals generally do not consciously engage in strategic actions to pass on their genes, or at least we need not assume that they do. Natural selection, in principle, does favour any behaviour of animals which generates above average reproductive success, as though the actors were consciously seeking a specific goal or result, in this case maximum inclusive fitness.

On the understanding that bisexual reproduction is indeed advantageous (Bernstein, Hopf, and Michod, Chapter 1) both sexes should cooperate in order to reproduce successfully. However, since the reproductive conditions and physiological constraints of both sexes are in general different, we should expect different reproductive strategies as well. Moreover, conflicts between such sex-specific reproductive strategies are predictable, and may sometimes even escalate to the famous 'battle of the sexes'. Anyhow, these conflicts must eventually end up in compromises in order to result in successful reproduction thus achieving the final goal of both sexes. Each sex, however, is expected to try to manipulate its counterpart for its own interests. Alternative strategies often develop which, by frequency dependent selection, may result in mixed evolutionarily stable strategies (ESS). Besides *intersexual* competition, *intra-*

sexual competition plays an important part in modelling reproductive strategies. For instance, infanticide can represent either male–male competition (e.g. in langurs, lions, etc.) or female–female competition (e.g. among prairie dogs). Observations of langur monkeys (*Presbytis entellus*) illustrate how infanticide can be the result of male–male competition for reproduction. Several langur populations breed in harems. There are few harem leader positions within the population, resulting in severe male–male competition for harems, and relatively short tenureships for each harem-possessing male. Under these conditions a male is able to increase his reproductive success by killing those infants sired by his predecessor which still hold their mothers in the state of lactational amenorrhea, thereby bringing those females quickly into estrus again. Field data (Sommer, 1987) strongly support Hrdy's (1974) hypothesis that male infanticide in langurs is an adaptive reproductive strategy evolved through sexual selection pressures on males. In a model considering female reproductive characteristics and the frequency distribution of male tenure lengths within the population, it can be shown that a mixed ESS may be established, stabilizing a fixed proportion of infanticidal and non-infanticidal 'take-overs' of harems (Hausfater, 1984). Langur females seldom succeed in their efforts to defend their infants against male infanticide. Thus, in this case males appear to be the winners of an intersexual reproductive conflict. On the other hand, Linsenmair (Chapter 2), referring to the terrestrial isopod *Hemilepistus*, gives a clear example where 'males are manipulated to their utmost disadvantage but to the great advantage of the manipulating females'.

The basic reason for intersexual reproductive conflicts is apparently rooted in the fundamental difference between the sexes. One sex, the female, specializes in the production of a smaller number of relatively large gametes, called eggs, the other sex, the male, produces a larger number of relatively small gametes, called sperm. The sex that produces the smaller number of larger gametes, the female, has much more to invest in each of those eggs than the male has to invest in its sperm, and this results in the basic asymmetry of interests which is particularly apparent in higher vertebrates but also, as Linsenmair (Chapter 2) demonstrates, in some highly specialized invertebrates. For instance, in mammals, the females during pregnancy have to nourish their unborn offspring. Not only do mothers run the risks associated with pregnancy and childbirth, but after delivery they have to produce milk during the period of lactation and to exercise various, sometimes very expensive, sorts of caretaking as well. Males, on the other hand, can easily produce sperm in abundance which after each ejaculation can be readily replaced within a very short time and males are usually able to move away from the consequences of their copulations. 'At minimum', as Barash (1977) puts it, 'males invest a little bit of time and sperm; females may have to live for weeks, months, or a lifetime with the consequences of their reproductive decisions'. Thus, females are usually the limiting reproductive resource for males who have to compete for access, and the variance of reproductive success is usually greater in males than in females. These basic differences necessarily result in diverse sex-specific reproductive strategies by means of natural selection. In general, females are expected to be the more choosy sex, whereas males are expected to be the more opportunistic sex.

These reflections may present a conceptual framework for looking at the competing sexual and reproductive strategies of both sexes regarding (a) the activities and decisions associated with mating (mating strategies) and (b) the activities and decisions associated with parental care (parental investment strategies).

Reproductive success requires optimal mating, and optimal mating is the consequence of advantageous strategic decisions. The potential reproducer has to decide when to mate and with whom to mate. Females have to decide whether to engage in immediate mating (e.g. copulating without prior court-ship) or whether to be coy (e.g. insisting on prolonged courtship before permitting copulations). Males may choose between being faithful (even willing to undertake a prolonged courtship) or being philanderers and desert-ing the female after fertilization. Females, in general, tend to be more choosy compared with males who tend to maximize the number of mates. This holds true for humans as well (Symons and Ellis, Chapter 8; Grammer, Chapter 9). Both sexes following different options: females, for instance, commonly look-ing for high status males, older than themselves, who can provide material and emotional security; males, in contrast, ordinarily seeming to decide on the basis of physical and sexual attractiveness and preferring younger females with higher reproductive value.

In many species both sexes dispose of alternative mating strategies. For instance, males may patiently consort with one female or try to sneak copulations (Kuester and Paul, Chapter 6) or even rape; females may, inten-tionally or not, attract and mate with several males in succession or may solicit only one male. There are various explanations for multiple matings of females, such as avoidance of harassment, formation of protective or parental care alliances with multiple males or simply the high energetic costs of resist-ing copulations. Mate guarding or harassment of sexual competitors are additional strategies to secure mating success (Linsenmair, Chapter 2; Kues-ter and Paul, Chapter 6), but mate guarding may serve other purposes as well (Lamprecht, Chapter 3).

Besides individual mate choice and mate management, animals as well as humans should decide on the type of mating system to reproduce in: e.g. would it be more advantageous to reproduce within a monogamous, a poly-gynous, polyandrous or promiscuous mating system? All these strategic decisions are closely related with physiological as well as with ecological constraints on the respective species. Under certain conditions intrasexual alliances may substantially enhance the personal and/or inclusive fitness of the animal concerned (Dunbar, Chapter 5).

Reproductive success frequently depends heavily on the form and intensity of parental investment. However, parental care is rarely equally distributed between the sexes. The intersexual conflict regarding the necessary investment for the common offspring has been described as the parental investment game: 'who gets left with the baby?' (Hammerstein and Parker, 1987). In cases where only one parent is required for a successful rearing of the offspring, one sex is expected to try to desert its mate. Physiological constraints usually decide which sex has to carry the burden of parental care. In fish, for example, where both males and females are usually equipped similarly, males or females may guard the offspring. In mammals, however, where the physiological conditions of males and females are extremely different, males often try to avoid the costs of parental care and to maximize their reproductive success by increasing the number of mates. This also applies to several other groups of animals, for instance to some terrestrial isopods (Linsenmair, Chapter 2). Among species where assistance of both mates is required to rear offspring successfully, environmental conditions – such as distribution, availability, and defensibility of limiting resources – determine the relative investment of the two sexes in their common offspring.

In those species requiring considerable parental investment of both mates in the rearing of offspring, mechanisms should have evolved that prevent individuals from wasting their time and energy on unrelated offspring. In general, this problem does not exist for females (except in cases of net parasitism); males, however, are more susceptible to this trap because they cannot be completely confident of being genetically related to the young they help to rear. This is the more important as females may use sophisticated tactics to deceive their male partner with respect to paternity. Manipulation of one sex by the other can frequently be observed, in animals as well as in humans.

A more complicated way of manipulating other conspecifics in one's own reproductive interest is to convert them into 'helpers at the nest'. This is most frequently done by reproductive suppression, either through parental manipulation (e.g. in eusocial insects) or by stress-induced reproductive suppression of kin or non-kin group members, in some cases for purely competitive reasons (Dunbar, Chapter 5). The mechanisms of reproductive suppression may be diverse and they have not yet been analysed sufficiently in most cases. This book contains examples of the 'helper at the nest' syndrome in animals (Rasa, Chapter 4) as well as in human society (Flinn, Chapter 12). It seems likely that 'infant transfer' as in langur monkeys (Sommer, Chapter 7) and 'babysitting' are possible preconditions to develop 'helper at the nest' systems. Maximizing inclusive fitness may be the ultimate purpose of helping close relatives to raise their offspring.

Another important strategic decision in reproductive effort is how to distribute limited resources among the offspring. Reproductive strategies of differential investment allocation are well known among numerous species. Sex ratio manipulation by sex biased infanticide is but one famous strategy for enhancing long-term reproductive success (Schiefenhövel, Chapter 10; Irwin, Chapter 13). Individual lifetime reproductive management, however, plays an important role in maximizing inclusive fitness (Voland and Engel, Chapter 11; Blurton Jones, Chapter 14).

As has become evident by means of meticulous investigations humans, like animals, act on average so as to maximize their reproductive success, even if females under certain conditions space births widely (Blurton Jones and Sibly, 1978; Blurton Jones, Chapter 14). Successful reproductive management through lifetime also requires an optimal social placement of children, which on average leads to parental fitness gains (Voland and Engel, Chapter 11). Evolutionary models of sexual and reproductive strategies are helpful for understanding patterns of human reproduction, and the results of careful research on animals allow for substantial predictions and testable hypotheses with regard to human reproductive behaviour.

Besides the analogies already mentioned (such as mating strategies and courtship, cooperative breeding systems, reproductive suppression and 'helpers at the nest', as well as sex ratio manipulation through infanticide) another example appears to be particularly instructive. Unlike most mammals, in which ovulation is advertised, in human females the time of ovulation is concealed. By this, human males are practically forced to copulate regularly with the same woman to ensure fertilization as well as guarding the female to secure paternity. These conditions, unquestionably, promote the social bond between the heterosexual partners and stimulate paternal investment, which, for several reasons, seems to have been extremely important during early hominid evolution. Interestingly enough, Linsenmair (Chapter 2) describes an analogy with the isopod genus *Hemilepistus*. Here, too, the female withholds important information about her reproductive state and by this forces the male to guard her in order to avoid being cuckolded. Males are manipulated to the advantage of females and their offspring. This, of course, is comparable to the human female strategy of concealed ovulation, at least in its social consequences. Males, as a result of natural selection, should be reluctant to waste time and energy on the rearing of unrelated offspring. Thus, it seems reasonable, that in several human societies, where female promiscuity is commonplace, the inheritance of property is through the sister of the male and not through his wife, since the children of the male's sisters are certainly genetically related to him, while at least some of his wife's children may be unrelated (Alexander, 1979).

The chapters of this book clearly demonstrate that animals as well as humans are skillful strategists with regard to their genetic reproduction, a capacity inevitably resulting from the continuous pressures of natural selection during evolution.

REFERENCES

Alexander, R.D. (1979) *Darwinism and Human Affairs*, University of Washington Press, Seattle and London.

Barash, D.P. (1977) *Sociobiology and Behavior*, Elsevier, New York.

Blurton Jones, N. and Sibly, R.M. (1978) Testing adaptiveness of culturally determined behaviour: do Bushman women maximize their reproductive success by spacing births widely and foraging seldom? In *Human Behaviour and Adaptation* (eds. N. Blurton Jones and V. Reynolds), Taylor and Francis, London, pp. 135–57.

Hammerstein, P. and Parker, G.A. (1987) Sexual selection: games between the sexes. In *Sexual Selection: Testing the Alternatives* (eds D.W. Bradbury and M.B. Anderson), John Wiley, Chichester, pp. 119–42.

Hausfater, G. (1984) Infanticide in langurs: strategies, counterstrategies, and parameter values. In *Infanticide: Comparative and Evolutionary Perspectives*, (eds G. Hausfater and S.B. Hirdy), Aldine, New York, pp. 257–81.

Hardy, S.B. (1974) Male–male competition and infanticide among the langurs (*Presbytis entellus*) of Abu, Rajasthan. *Folia primatol.*, **22**, 19–58.

Sommer, V. (1987) Infanticide among free-ranging langurs (*Presbytis entellus*) at Jodhpur (Rajasthan/India): recent observations and a reconsideration of hypotheses, *Primates*, **28**, 163–97.

PART ONE

Eco-ethological aspects of sexuality and reproduction in animal populations

CHAPTER ONE

The evolution of sex: DNA repair hypothesis

Harris Bernstein, *Frederic A. Hopf*
and Richard E. Michod

1.1 SEXUAL REPRODUCTION IS A COSTLY PROCESS

The large investment of resources devoted to sexual reproduction by most organisms is impressive. A large investment implies that the adaptive benefit must be comparably great. Yet the benefit of sexual reproduction (sex) is widely regarded by evolutionary biologists as one of the major unsolved problems of biology.

The short-term costs of sex involve the cost of males (Williams, 1975; Maynard Smith, 1978), high recombinational load (Shields, 1982), lower genetic relatedness between parent and offspring (Williams, 1980; Uyenoyama, 1984, 1985) and the cost of mating (Bernstein *et al.*, 1985d; Hopf and Hopf, 1985). These costs either stem directly from outcrossing, as in the case of the costs of mating and males, or are enhanced by outcrossing, as in the case of recombinational load and genetic relatedness (Uyenoyama, 1988). Therefore, any fundamental explanation of outcrossing, to be plausible, must provide a benefit that is large enough to account for its costs.

1.2 TRADITIONAL IDEAS ON THE BENEFIT OF SEX

For most of the past 50 years, biologists generally regarded the benefit of sex as a problem that had been resolved. The traditional explanation is based on an obvious feature of sex. As a general rule, progeny resulting from sexual reproduction are genetically different from their parents, whereas in asexual reproduction they are the same as their sole parent. Sexually produced progeny not only differ from their parents, but also from each other (except

The order of authors is strictly alphabetical and is not intended to imply seniority.

for identical twins). Thus sexual reproduction produces new genetic variants at each generation. By the traditional view, genetic variation is advantageous because it facilitates adaptation.

In the past 15 years, much theoretical work has been done to define the genetic and ecological conditions under which sexually produced genetic variation is advantageous. This work, rather than sustaining the variation explanation, has shown that it operates only under limited conditions. The realization that this longstanding explanation has serious problems in accounting for the ubiquity of sex in nature has led to the now widely expressed opinion, mentioned above, that the function of sex is a major unsolved problem in biology.

1.3 THE DNA REPAIR HYPOTHESIS

A new explanation of sex based on another obvious feature, which we consider to be more fundamental than variation, has recently been proposed (Bernstein, 1979; Bernstein, Byers and Michod, 1981; Bernstein *et al.*, 1984, 1985a,b,c; Bernstein, Hopf, and Michod, 1987, 1988). Progeny are not only genetically different from their parents, they are also young. There are two aspects to youth. The first aspect, which does not play a significant role in our explanation, relates to the process of development from fertilized egg to the mature organism. Development is thought to occur by a sequential switching on, or off, of genes. The developmental clock needs to be reset at each generation, and this presumably occurs when germ cells (egg and sperm) are formed.

The second aspect of youth is that, whereas parents age, cells of the germ line do not, so that progeny do not reflect the aging of their parents. We argue that the lack of aging of the germ line results mainly from repair of the genetic material by meiotic recombination during formation of germ cells. Thus our basic hypothesis is that the primary function of sex is to repair the genetic material of the germ line. We refer to this as the repair hypothesis to contrast it to the traditional variation hypothesis. We will now explain the repair hypothesis starting from first principles.

When information of any kind is transmitted, it is subject to disruptions due to random influences which are collectively termed noise. This applies to the genetic material as well. In DNA and RNA genomes (RNA genomes occur in some viruses), noise occurs as a result of mutation and damage. These present a serious problem for survival, and early in evolution strategies probably evolved to reduce damage and mutation.

To facilitate further discussion we need to compare damage and mutation. Damages are abnormalities in the structure of DNA and RNA produced by reactive chemicals or radiation. Common examples are breaks in one or both strands of DNA, oxidatively altered nucleotides, and thymine dimers caused by ultraviolet (UV). Damages frequently block replication or prevent gene expression, and a single unrepaired damage may be lethal to a cell. Damages are not replicated and thus are not inherited. Generally damages can be recognized by enzymes which repair the genetic material. Mutations, in contrast to damages, involve changes in the coding sequence of standard bases in DNA or RNA genomes. They frequently lead to reduced or abnormal gene expression and thus often reduce fitness. Mutations are replicated and thus inherited. They are not detectable by enzymes and thus are not repaired. Mutation results from replication errors, either during genome replication or during the synthesis associated with repair of damages. Although mutations are most often disadvantageous, they can with low frequency create new adaptive information and thus promote evolutionary advance. Despite the evolutionary importance of beneficial mutations, the machinery of replication has been selected to be highly accurate to avoid the predominantly deleterious mutations. Mutations are often eliminated from a population of organisms by natural selection against less fit individuals. The expression of deleterious mutations may also be masked by expression of a second functional copy of the gene in the same cell. The mutual masking of mutations in two genomes is called complementation.

Sexual reproduction has two fundamental aspects, whose benefits we will explain by the repair hypothesis. The first one is recombination, mentioned briefly above. To discuss this properly we need to give some background information. The genome of most organisms is divided into a set of distinct chromosomes, each chromosome being comprised of a long DNA molecule plus smaller molecules which govern gene expression and mechanical properties of the chromosome. Each diploid cell contains two sets of homologous chromosomes with equivalent DNA sequences. Homologous chromosomes are not necessarily identical since they may differ by mutations that are newly arisen or acquired by descent. Recombination refers to the process of breakage and exchange of segments of DNA between two homologous chromosomes present in the same cell. Recombination is widely regarded as a fundamental attribute of sex. In multicellular organisms a key stage of meiosis, the process by which germ cells are formed, is designed to promote recombination. We consider that the function of recombination is to repair genome damage in the germ line. The second fundamental aspect of sex is outcrossing. This means that the two homologous chromosomes which engage in recombination come from separate parents. This contrasts with asexual reproduction in which the two chromosomes present in any cell are from the same parent. We think the benefit of outcrossing results from complementation, the mutual masking of mutations in homologous chromosomes. In summary, we consider that the two fundamental aspects of sex are recombination and outcrossing, and these provide the respective benefits of repair of genome damage, and masking of deleterious mutations.

1.4 DNA DAMAGE

DNA damages occur in a variety of ways. Sometimes, the damaging agent is from an external source. In humans, UV irradiation from the sun directly damages DNA, causing skin cancer. Other external sources are chemicals in our environment, particularly in our diets.

Probably the most important sources of damage in germ cell DNA are processes that occur naturally inside the cell. A class of very reactive molecules, oxidative free radicals, are produced naturally as a byproduct of energy transformations in a metabolizing cell. It has been estimated in humans that several hundred damages occur in the DNA of an average cell per day due to the action of these free radicals (Cathcart *et al.*, 1984; Ames *et al.*, 1985). Other internal sources of DNA damage are also known. In general, DNA damages can be classified into two types, those that affect only one strand of the DNA and those that affect both strands. Most damages are of the single-strand type, and there are many different kinds. Double-strand damages are less frequent and are of only a few kinds. Examples are breaks in the double helix in which both strands are severed (double-strand breaks), and cross-links in which the two strands of DNA cannot pull apart because they are joined by a covalent bond. Often, if there is a damage in only one strand, when this strand is replicated a gap occurs in the newly formed strand opposite the damage. This condition is now a double-strand damage since information is missing at the same position in both strands of the DNA.

1.5 DNA REPAIR

DNA damages are deleterious, and organisms have evolved ways of repairing them. Since damages cause loss of information from the DNA sequence, repair processes depend on obtaining the information from another intact sequence. For single-strand damages in double-stranded DNA this is relatively easy. The damage is often detected by a special enzyme designed to recognize that particular type of damage. The enzyme, acting in cooperation with other enzymes, then cuts out the damage along with a stretch of the strand to either side. This leaves a gap in one strand, but the other undamaged strand is not affected. Therefore, the gap can be filled in by copying the information from the intact partner strand. This process is called excision repair.

Double-strand damages cannot be repaired by excision repair because the information at the damaged site is entirely lost to the DNA. Repair can occur, however, if a second DNA molecule is nearby and information is exchanged between the two DNA molecules. Thus, in this type of repair, DNA with an intact sequence donates an undamaged single-stranded section to the damaged chromosome to permit replacement of its lost information. This process is referred to as recombinational repair.

6

1.6 RECOMBINATIONAL REPAIR IS EFFICIENT AND PREVALENT IN NATURE

Much of what is known about the molecular aspects of recombinational repair comes from research on simple organisms such as yeast, bacteria and viruses (Bernstein *et al.*, 1987). These studies show that recombinational repair is effective against many types of DNA damages. Of special relevance are studies showing impressive effectiveness against double-strand damages. In yeast, recombinational repair of double-strand breaks is about 97% efficient, and of crosslinks about 99% efficient (Resnick and Martin, 1976; Magana-Schwencke *et al.*, 1982). Although much less research has been done on recombinational repair in multicellular organisms, indirect evidence suggests that the process is similar.

1.7 MEIOSIS

In meiosis a diploid cell produces four haploid daughter cells. First each chromosome in the diploid cell replicates producing four sets of chromosomes. Next, in a process unique to meiosis, the four homologous chromosomes line up with each other. Recombination then occurs between pairs of homologous chromosomes. When this is completed there are two sequential cell divisions producing four haploid cells. These are the germ cells. Although the events of germ cell formation preceding and following meiosis differ from organism to organism, meiosis is a consistent feature. According to the repair hypothesis meiosis is specially designed to promote recombination repair of germ line DNA.

1.8 MEIOTIC RECOMBINATION AND DNA REPAIR

The problem of how recombination occurs is one of the classical problems in genetic research and a great deal of detailed information has been accumulated about the process. The purpose of this work has been to achieve a model of recombination which explains, as simply as possible, the accumulated experimental data. The most recent model to gain widespread acceptance is referred to as the double-strand break repair model (Szostak *et al.*, 1983). Although it is uncertain that this model will ultimately prove correct, its central features are solidly based on experimental observation. It should be noted that no evolutionary assumptions about the benefit of recombination were made in designing the model. According to this model the first step in recombination is a double-strand break in one chromosome which is extended to a gap. This chromosome then receives genetic information to fill the gap from the homologous chromosome with which it is intimately paired. The succeeding process of exchange is explained in detail by the model, but it is not necessary to review these details here. The most significant aspect from the perspective of the repair hypothesis is the first step. The initial formation of a double-strand gap is what one would expect if recombination were designed to remove any double-strand damages in a simple direct fashion.

1.9 THE ORIGIN AND EARLY EVOLUTION OF SEX

Since genome damage is likely to have been a significant problem at a very early stage in evolution, the repair hypothesis implies that sex also should have evolved very early. We think it likely that protocells, the most primitive single-celled organisms, evolved a simple form of sex. Protocells were probably haploid as are present day bacteria. Several authorities on early evolution (Eigen and Schuster, 1979; Eigen *et al.*, 1981; Woese, 1983) have suggested that the genomes of early protocells were composed of single-stranded RNA, and that individual genes were coded by separate RNA segments, rather than being linked end-to-end as in DNA. One of the simplest living organisms, the influenza virus, has a genome organization similar to that postulated for protocells. The genome of the influenza virus is composed of eight separate segments of single-stranded RNA and six of these code for only one gene, whereas the others code for two or three genes. When an influenza virus infects a cell, each of its RNA segments replicates repeatedly, and viable progeny viruses contain replicas of the original set of eight RNA segments. If one of the RNA segments of the parental virus is damaged, its replication or expression may be blocked and no viable progeny viruses are produced.

Influenza viruses can undergo a primitive form of sex. If two undamaged viruses infect a cell, each progeny virus will have a complete set of segments, but each set is likely to contain a mixture of genes from both parents. This exchange of genes is a simple form of recombination. If the two infecting viruses both have damaged genomes, viable progeny may still be formed as long as there is at least one undamaged copy of each segment among the two infecting genomes. Thus two viruses which would have been inviable alone due to damage can mutually reactivate each other. We regard this reactivation of influenza viruses as an appropriate model for sex in protocells (Bernstein *et al.*, 1984). If protocells with damaged RNA segments fuse with each other, mutual reactivation can occur. The progeny following fusion would again be undamaged haploid protocells. We think the cycle from haploidy, to diploidy with recombination, and then back to haploidy, is the sexual cycle in its most primitive form. Its benefit is to allow damaged protocells that are inviable alone to survive by cooperating with each other.

A problem with having a segmented genome is ensuring that, upon replication, a complete set of genes is passed to each progeny virus. In fact, influenza viruses often form defective progeny with incomplete sets of genes because of unreliable segregation. As protocells evolved increasing numbers of genes the segregation problem must have become progressively more acute. One straightforward way of solving this problem is linking genes end-to-end. For this reason, we think that segmented RNA genomes evolved into the typical DNA genomes where genes are linked end-to-end in a long chain. However, when genes are linked in this way, recombination to form undamaged genomes cannot occur merely by assortment of undamaged replicas. Special enzymes had to evolve to allow the splicing of undamaged segments of DNA to replace damaged segments. Thus enzyme-mediated recombinational repair emerged. As we discussed in the previous section meiosis seems designed to promote this process. In conclusion, we think sex had a continuous evolutionary development from the earliest protocells to current multicellular organisms based on the benefit of recombinational repair.

1.10 THE EMERGENCE OF DIPLOID ORGANISMS

In the previous section we postulated that protocells were haploid, but periodically fused to form transient diploids in order to reactivate damaged genomes. One might wonder if a better solution would be to maintain a permanent state of diploidy. We think the reason this option is not used by simple organisms is that it is not cost effective because of the burden of making an extra genome per cell. We have shown (Bernstein *et al.*, 1984) that under a wide range of conditions the best strategy is one that combines the advantage of the rapid rate of asexual reproduction of haploids with the option of periodically fusing, when damaged, to restore reproductive ability.

As haploid organisms evolved, acquiring more adaptations, their genome expanded to encode additional genes. As the number of genes increased, there was a corresponding increase in vulnerability to mutation. It is likely that this problem was dealt with by improving the accuracy of the enzymatic machinery that replicates DNA. This inference is based on evidence in extant haploid organisms that accuracy of replication increases in direct proportion to the size of the genome. We think, however, that a stage of evolution was eventually reached where an alternative strategy for coping with mutation became more cost effective than further improvement of the replicative machinery.

This option is complementation. As noted previously complementation is the mutual masking of mutations when two genomes share the same cell. Complementation can be exploited if an organism becomes a predominant diploid, with the haploid cells (germ cells) exposed only briefly to environmental selection. In outcrossing sexual organisms, the two genomes brought together in a common cell are from independent organisms and are therefore unlikely to have mutations in common genes. Once an organism becomes committed to predominant diploidy, it runs into a new problem. Because deleterious mutations are masked they are no longer efficiently weeded out by natural selection. Thus a lot of mutations build up in the diploid line. Eventually an equilibrium is reached where the expression of deleterious mutations in the diploid line is as frequent as it was in the original haploid line. Although the advantage of diploidy is transient, reversion back to haploidy is blocked because this would lead to immediate expression of all deleterious mutations accumulated in the diploid.

1.11 WHY HAVE TWO PARENTS (OUTCROSSING) WHEN ONE IS SUFFICIENT (PARTHENOGENESIS)?

We have briefly reviewed evidence for the central role of repair in recombination. Recombinational repair requires the presence of two chromosomes in a common cytoplasm. There are two potential strategies for obtaining these two chromosomes: (a) a closed system strategy where they both come from the same parent in the previous generation; or (b) an open system strategy where they come from different parents. If recombinational repair were the only selective advantage of sexual reproduction, the most effective strategy would be a closed system such as a self-fertilization or automixis (uniparental production of eggs through ordinary meiosis followed by some internal process for restoring diploidy). This would avoid the major costs of sex enumerated above. Thus, we need to explain why the most common strategy is outcrossing sex, an open system.

We have argued that outcrossing arose in primitive protocells because of the cost of maintaining more than one set of chromosomes within a protocell. These costs stem from the simple fact that, for a cell to replicate, all its chromosomes must replicate. For simple protocells, the costs of replicating the genetic material are a major component of its overall resource and time budget. Consequently, for simple cells, the costs of replicating a diploid cell are approximately twice that of replicating a haploid cell. Later, with the evolution of multicellular life, the costs of replicating the genetic material became a smaller portion of the total resource budget. As these costs of diploidy decreased, diploidy eventually emerged as the predominant stage of the life cycle because of the advantage of masking deleterious recessive alleles through complementation (section 1.10). We now argue that the advantage of complementation also explains the maintenance of outcrossing in reproductive systems that have recombination (Bernstein *et al.*, 1985b; Lande and Schemske, 1985; Hopf *et al.*, 1988).

1.11.1 Reproductive system and recessive mutations

Mutations result from errors of replication. Improvements in replication accuracy have costs which include increased energy use (Hopfield, 1974), additional gene products (Alberts *et al.*, 1980) or slower replication (Gillen and Nossal, 1976). Thus there are probably cost-effectiveness barriers to indefinite improvement in accuracy, and a finite spontaneous mutation rate is likely to be an intrinsic property of genome replication. Felsenstein (1974) has noted that in the short run deleterious mutations affect fitness much more than beneficial ones because of their much higher rate of occurrence. Haldane (1937) argued that in a population at equilibrium deleterious mutations are removed by selection at the same rate that they arise by mutation. He demonstrated that the average survivorship at equilibrium is solely a function of μ, the rate of deleterious mutation per haploid genome, and is not affected by how harmful the individual mutations are.

Table 1.1 lists the reproductive systems of diploid organisms. We have found by a generalization of Haldane's argument that in all of these diploid reproductive systems survivorship due to expression of deleterious recessive alleles is $e^{-\mu}$ at equilibrium (Hopf *et al.*, 1988). Haploid and haplodiploid organisms (not listed in Table 1.1) require special consideration and are discussed elsewhere (Bernstein *et al.*, 1985b, 1987). Since the effect of μ on survivorship is the same for all reproductive systems, μ should not differ among these systems. Consequently, we assume μ to be fixed, and consider below the effect of switching from one reproductive system to another among those listed in Table 1.1.

First, we estimate a value for μ. Data obtained by Mukai *et al.* (1972) indicate that, in *Drosophila*, $\mu = 0.3$ for lethal and non-lethal mutations

Table 1.1
Classification of diploid reproductive systems[*]

Reproductive system	Masking ability at equilibrium	Recombinational repair	Source of homologous chromosome
Automixis	Low ($\approx 2\mu$)	Yes	Self
Selfing	Low ($\approx 2\mu$)	Yes	Self
Outcrossing[†]	Intermediate $[\approx \sqrt{n\mu}]$ [‡]	Yes	Another individual
Panmixia	Intermediate ($\approx n$)	Yes	Another individual
Endomitosis	High ($\approx n$)	Limited[§]	Self
Apoximis	High ($\approx n$)	No	Not applicable
Vegetative	High ($\approx n$)	No	Not applicable

[*] See Hopf, Michod and Sanderson (1988) for supporting details.
[†] With some mating between relatives as occurs in nature.
[‡] n denotes the number of functional genes per genome, which, in higher organisms, is approximately 40,000.
[§] See text (section 1.11.2).

11

combined. This gives an equilibrium survivorship $e^{-\mu} \approx 0.7$, which is neither so low as to be devastating or so close to 1.0 (no reduction in survivorship) as to indicate a negligible effect of mutational load. If the mutational load were negligible we think it would lead to an increase in the number of genes, n. Since μ should be proportional to n, selection for increased n would also increase μ. Therefore, we consider 0.7 to be an approximate general value for $e^{-\mu}$, and 0.3 a general value of μ.

The second column of Table 1.1 lists the number of lethal mutations accumulated at equilibrium in the different reproductive systems. The differences between these systems in the number of accumulated recessive mutations are substantial. Those reproductive systems effective at masking recessive mutations accumulate many mutations. Those which are ineffective at masking accumulate few. These different numbers of mutations create transient selection pressures on mutant individuals which have breeding systems different from that of the majority equilibrium population.

1.11.2 Selection of reproductive system

The effect of mutational load on survivorship makes all systems equally competitive at equilibrium. However, there is a transient selective advantage to moving towards greater masking (downwards in Table 1.1); and a transient disadvantage to moving (upwards) towards diminished masking. There is a similar transient advantage in the shift from haploidy to diploidy. To illustrate the transient advantages and disadvantages of shifting between reproductive systems, we first consider a population fixed for selfing. Although individuals in this population will have accumulated few mutations at equilibrium (Table 1.1), new ones occur each generation at a frequency μ and a new mutant outcrosser will mask these mutations in its offspring. There is nearly complete complementation in an outcross because deleterious alleles in the two partners are statistically unrelated to each other. Thus, the fitness of the outcrosser is not initially reduced at all by mutation load, and it has a survivorship of unity due to mutation load. However, the new outcrossers must pay the costs of outcrossing sex described above. Let C be the factor by which fitness is reduced due to these costs. Thus C is the ratio of the fitness of the selfer, taken as unity, to the fitness of the new outcrosser (<1), so that $C>1$. We can now compare the cost C of the shift to outcrossing with the benefit of the shift. The benefit is given by $1/e^{-\mu} = e^{\mu}$, which is the factor by which fitness is increased due to the masking of deleterious recessive alleles when the shift occurs. If the cost is less than the benefit ($C < e^{\mu}$) then a gene for outcrossing should expand in the population. Taking μ to have a general value of 0.3 as discussed above, the benefit e^{μ} has a value of 1.4. Thus the shift to outcrossing should occur even if the costs are fairly large.

As outcrossing becomes fixed in a population, deleterious recessive alleles increase. Eventually the outcrosser has an equilibrium fitness that is reduced both by the costs of outcrossing and the mutational load. That is, the long-term effect of the transition to outcrossing is a net reduction of individual fitness. A successful shift back to selfing is inhibited, however, by the drastic consequences of unmasking the many accumulated deleterious recessive mutations. The shift from outcrossing to selfing will only succeed when the costs of outcrossing became very large. Among the first four reproductive systems listed in Table 1.1, which are those with full recombinational repair, outcrossing is favoured. An intermediate level of outcrossing may be preferred over panmixia because of the need to preserve coadapted gene complexes (Shields, 1982).

The transient advantage of complementation favours asexual systems in which the diploid maternal genome is passed down intact from mother to daughter, since this gives maximal masking of deleterious recessive alleles. However, the absence of recombinational repair in these asexual systems, as listed in column 3 of Table 1.1, also needs to be considered in evaluating the success of a shift from outcrossing to asexual reproduction. Apomixis, a parthenogenetic system, is characterized by suppression of meiosis and its replacement by a single mitotic maturation division. In vegetative reproduction, meiosis is also absent. In a shift from outcrossing to apomixis or vegetative reproduction, repair of double-strand damages is probably largely abandoned, and thus these strategies are costly. However, the effect of DNA damage may be overcome in these cases by cellular selection resulting from the death of damaged cells and the replication of undamaged cells. Cellular selection is effective when damages are infrequent and/or growth is rapid, but not otherwise.

Endomitosis, another parthenogenetic system, is characterized by two sequential premeiotic chromosome replications followed by an apparently normal meiosis to produce diploid eggs (Maynard Smith, 1988). Also, at the four chromatid stage, pairing is between chromatids derived from only one initial chromosome (Cuellar, 1971; White, 1973; Cole, 1984). This reproductive system is used, for example, by whiptail lizards common in the deserts of the south western United States. Because there is no recombination between the non-sister homologues, the maternal genome should be passed on intact to daughters. If recombination between non-sister homologues were allowed, the transient advantage would be lost because of the immediate expression of accumulated recessives. Endomitosis might seem to be an ideal strategy since it reaps the benefit of meiotic repair while avoiding the expression of deleterious recessive alleles. However, double-strand damages occurring before the first premeiotic replication cannot be repaired by endomitosis, as all chromatid pairing partners are derived from the same chromosome and there is no intact template corresponding to the damaged site. This problem does not arise in conventional meiosis, because recombination is between non-sister chromatids. Hence, if double-strand damages are common before premeiotic replication, endomitosis is an unsatisfactory option.

In conclusion, shifting from outcrossing to any of the reproductive systems in Table 1.1 results in an immediate reduction in fitness. This reduction is transient for selfing and automixis, since the few progeny that survive the unmasking of recessive mutations will be those with statistically fewer mutations. In contrast, the reduction in fitness upon shifting from outcrossing to either apomixis, endomitosis or vegetative reproduction is permanent since these latter systems have reduced capacity for recombinational repair.

Lynch (1984) has summarized the literature indicating that parthenogens often have lower reproductive rates than their sexual relatives, frequently less than 50%. Newly arisen parthenogens seem to experience a greater reduction in fecundity than established ones. The costs of parthenogenesis predicted by theory and supported by the literature imply that parthenogens would only be competitive in situations where the costs of outcrossing are larger than the costs of parthenogenesis. One common situation of this type is in newly created natural habitats (such as produced by floods and fires) where finding a mate is difficult. Evidence reviewed by Bernstein *et al.* (1985d) supports the generalization that parthenogens are favoured where the costs of finding a mate are high.

1.12 THE TRADITIONAL VIEW OF THE ADVANTAGE OF SEX: THE VARIATION HYPOTHESIS

In section 1.2 we mentioned the traditional view that the benefit of sex arises from the production of recombinational variation. Our view is that recombinational variation is a byproduct of recombinational repair just as mutation, the other principal source of inheritable variation, is a byproduct of DNA replication. Inheritable variation is essential for evolutionary advance. However, both mutations and recombination are random processes, so that mutational and recombinational variants are much more likely to be deleterious than beneficial. The enzymatic machinery that replicates DNA has been selected to be highly accurate so that mutation rates are extremely low. We can ask if the machinery of meiosis, like that of replication, has been selected to reduce random variation. Two aspects of meiosis indicate that this is the case (Bernstein *et al.*, 1988). First, about two-thirds of recombination events involve exchange of only a short segment of DNA about the length of a gene or less. Only about one-third involve exchange of the flanking portions of the chromosomes which would maximize recombinational variation. If variation is the primary purpose of meiosis we would expect all exchanges to be of the latter type. An examination of the molecular mechanism of meiotic recombination indicates that it involves a complex step designed to reduce rather than maximize, recombinational variation.

The second aspect of meiosis indicating that it has been selected to reduce recombination is premeiotic replication. This gives rise to two identical copies (sister chromosomes) of each original chromosome. There does not appear to be a bias concerning which two chromosomes come together in a recombinational repair event, so that exchanges between sister chromosomes are prevalent even though they produce no recombinational variation. Thus premeiotic replication is unexpected and puzzling on the variation hypothesis. On the repair hypothesis there is an advantage to premeiotic replication if the DNA has single-strand damages, since these are converted into a form that can initiate the double-strand break repair mechanisms of recombination, discussed in section 1.8.

1.13 WHY IS THE GERM LINE POTENTIALLY IMMORTAL WHILE THE SOMATIC LINE INEVITABLY AGES AND DIES?

If, as we have proposed here, the principal function of meiosis is to repair damage in germ line DNA, then how do somatic cells, which lack this process, cope with damage? Somatic cells possess the capacity for other forms of repair, principally excision repair. However, there is much less recombinational repair in somatic cells than in germ cells. We infer that DNA damages, and particularly double-strand damages, are a frequent cause of somatic cell death. In a tissue where cells are able to proliferate, a dead cell may simply be replaced by replication of an undamaged one. However, some major cell types lack proliferative capacity, e.g. nerve cells. Thus loss of neuronal function due to accumulation of DNA damage may be a primary cause of aging and mortality in humans (Gensler *et al.*, 1987). In contrast, the potential immortality of the germ line, we think, is due to efficient recombinational repair during meiosis.

1.14 CONCLUSIONS

Sexual reproduction has been regarded traditionally as an adaptation for producing genetic variation through allelic recombination. Serious difficulties with this explanation have led many workers to conclude that the benefit of sex is a major unsolved problem in evolutionary biology. In our view, the two fundamental aspects of sex are recombination and outcrossing. These, we think, are adaptive responses to the two major sources of noise in transmitting genetic information, DNA damage and replication errors. We refer to this idea as the repair hypothesis to distinguish it from the traditional variation hypothesis. In the repair hypothesis, recombination is an adaptation for repairing damaged DNA. In replacing damaged sites, recombination produces a form of informational noise, allelic recombination, as a byproduct. Recombinational repair is the only repair process known which can overcome

15

double-strand damages in DNA, and such damages appear to be common in nature. Recombinational repair is prevalent from the simplest to the most complex organisms. It is effective against many different types of DNA damages, and in particular is highly efficient in removing double-strand damages. Current understanding of the mechanism of recombination during meiosis suggests that meiosis is designed for repairing DNA.

The evolution of sex can be viewed as a continuum from the repair hypothesis. Sex is presumed to have arisen in primitive RNA-containing protocells. We think that the earliest form of sex was similar to that of recombinational repair in extant segmented single-stranded RNA viruses. In such simple organisms recombinational repair occurs by nonenzymatic reassortment of replicas of undamaged RNA segments. As genome information increased during evolution, genes became attached end-to-end in long DNA molecules and recombinational repair evolved into enzyme-mediated breakage and exchange. Increased genome information also leads to increased vulnerability to mutation. The diploid stage of the sexual cycle, which was at first transient, became the predominant stage in some lines because it allows complementation, the masking of deleterious recessive mutations. Outcrossing, the second fundamental aspect of sex is also maintained in these lines of descent by the advantage of masking mutation. However, outcrossing can be abandoned in favor of parthenogenesis or selfing under conditions where the costs of mating are very high.

In multicellular organisms, the potential immortality of the germ line may be due to recombinational repair during meiosis, whereas the apparent inevitable aging of the organism may be due to accumulated unrepaired DNA damages in the somatic line.

ACKNOWLEDGEMENTS

This work was supported by grants NIH GM27219 (HB), NIH RO1 HD19949 (REM), NIH HD00583 (REM), and NIH GM 36410.

REFERENCES

Alberts, B.M., Barry, J., Bedinger, P., Burke, R.L., Hibner, U., Liu, C.-C. and Sheridan, R. (1980) Studies of replication mechanisms with the T4 bacteriophage *in vitro* system. In: *Mechanistic Studies of DNA Replication and Genetic Recombination*, (eds B. Alberts and C.F. Fox), ICN–UCLA Symposia on Molecular and Cellular Biology, Academic Press, New York, pp. 449–74.

Ames, B.N., Saul, R.L., Schwiers, E., Adelman, R. and Cathcart, R. (1985) Oxidative DNA damage as related to cancer and aging: assay of thymine glycol, thymidine glycol, and hydroxy-methyluracil in human and rat urine. In: *Molecular Biology of Aging: Gene Stability and Gene Expression* (eds R.S. Sohal, L.S. Birnbaum and R.G. Cutter), Raven Press, New York, pp. 137–44.

References

Bernstein, C. (1979) Why are babies young? Meiosis may prevent aging of the germ line. *Perspect. Biol. Med.*, **22**, 539–44.

Bernstein, H., Byerly, G.S. and Michod, R.E. (1981) Evolution of sexual reproduction: importance of DNA repair, complementation and variation. *Am. Natur.*, **117**, 537–49.

Bernstein, H., Byerly, H.C., Hopf, F.A. and Michod, R.E. (1984) Origin of sex. *J. Theor. Biol.*, **110**, 323–51.

Bernstein, H., Byerly, H.C., Hopf, F.A. and Michod, R.E. (1985a) The evolutionary role of recombinational repair and sex. *Int. Rev. Cytol.*, **96**, 1–28.

Bernstein, H., Byerly, H.C., Hopf, F.A. and Michod, R.E. (1985b) Genetic damage, mutation and the evolution of sex. *Science*, **229**, 1277–81.

Bernstein, H., Byerly, H., Hopf, F.A. and Michod, R.E. (1985c) DNA repair and complementation: the major factors in the origin and maintenance of sex. In: *Origin and Evolution of Sex* (ed. H.O. Halvorson), Alan R. Liss Inc., New York, pp. 29–45.

Bernstein, H., Byerly, H.C., Hopf, F.A. and Michod, R.E. (1985d) Sex and the emergence of species. *J. Theor. Biol.*, **117**, 665–90.

Bernstein, H., Hopf, F.A. and Michod, R.E. (1987) The molecular basis for the evolution of sex. *Advances in Genetics*, **24**, 323–70.

Bernstein, H., Hopf, F.A. and Michod, R.E. (1988) Is meiotic recombination an adaptation for repairing DNA, producing genetic variation, or both? In: *The Evolution of Sex: An Examination of Current Ideas* (eds B. Levin and R. Michod), Sinauer Associates, Sunderland, Mass., pp. 139–60.

Cathcart, R., Schwiers, E., Saul, R.L. and Ames, B.N. (1984) Thymine glycol and thymidine glycol in human and rat urine: a possible assay for oxidative DNA damage. *Proc. Natl. Acad. Sci. USA* **81**, 5633–7.

Cole, C.J. (1984) Unisexual lizards. *Sci. Amer.*, **250**, 94–100.

Cuellar, O. (1971) Reproduction and the mechanism of meiotic restitution in the parthenogenetic lizard *Cnemidophorus uniparens*. *J. Morph.*, **133**, 139–65.

Eigen, M. and Schuster, P. (1979) *The Hypercycle, a Principle of Natural Self-Organization*, Springer-Verlag, Berlin.

Eigen, M., Gardiner, W., Schuster, P. and Oswatitsch, P. (1981) The origin of genetic information. *Sci. Am.*, **244**, 88–118.

Felsenstein, J. (1974) The evolutionary advantage of recombination. *Genetics*, **78**, 737–56.

Gensler, H.L., Hall, J.D. and Bernstein, H. (1987). The DNA damage hypothesis of aging: importance of oxidative damage. In: *Reviews of Biological Research in Aging*, **3**, 427–41.

Gillen, F.D. and Nossal, N.G. (1976) Control of mutation frequency by bacteriophage T4 DNA polymerase I. The tsCB120 antimutator DNA polymerase is defective in strand displacement. *J. Biol. Chem.*, **251**, 5219–24.

Haldane, J.B.C. (1937) The effect of variation on fitness. *Am. Natur.*, **71**, 337–49.

Hopf, F.A. and Hopf, F.W. (1985) The role of the Allee effect on species packing. *Theor. Popl. Biol.*, **27**, 27–50.

Hopf, F.A., Michod, R.A. and Sanderson, M. (1988) On the effect of reproductive systems on mutation load and the number of deleterious mutations. *Theor. Pop. Biol.*, **33**, 243–65.

17

Hopfield, J.J. (1974) Kinetic proofreading: A new mechanism for reducing errors in biosynthetic processes requiring high specificity. *Proc. Natl. Acad. Sci. USA*, **77**, 4135–9.

Lande, R. and Schemske, D.W. (1985) The evolution of self-fertilization and inbreeding depression in plants: I. Genetic models. *Evolution*, **39**, 24–40.

Lynch, M. (1984) Destabilizing hybridization, general purpose genotypes and geographic parthenogenesis. *Q. Rev. Biol.*, **59**, 257–90.

Magana-Schwencke, N., Henriques, J-A.P., Chanet, R. and Moustacchi, E. (1982) The fate of 8-methoxypsoralen photoinduced crosslinks in nuclear and mitochondrial yeast DNA: comparison of wild-type and repair deficient strains. *Proc. Natl. Acad. Sci. USA*, **79**, 1722–6.

Maynard Smith, J. (1978) *The Evolution of Sex*, Cambridge University Press, Cambridge.

Maynard Smith, J. (1988) The evolution of recombination. In: *The Evolution of Sex: An Examination of Current Ideas* (eds B. Levin and R. Michod), Sinauer Associates, Sunderland, Mass., pp. 106–25.

Mukai, T., Chigusa, S.I., Mettler, L.E. and Crow, J.F. (1972) Mutation rate and dominance of genes affecting viability in *Drosophila melanogaster*. *Genetics*, **72**, 335–55.

Resnick, M.A. and Martin, P. (1976) The repair of double-strand breaks in the nuclear DNA of *Saccharomyces cerevisiae* and its genetic control. *Mol. Gen. Genet.*, **143**, 119–29.

Shields, W.M. (1982) *Philopatry, Inbreeding and the Evolution of Sex*. State University of New York Press, Albany.

Szostak, J.W., Orr-Wearer, T.L., Rothstein, R.J. and Stahl, F.W. (1983) The double-strand break repair model for recombination. *Cell*, **33**, 25–35.

Uyenoyama, M.K. (1984) On the evolution of parthenogenesis: a genetic representation of the 'cost of meiosis'. *Evolution*, **38**, 87–102.

Uyenoyama, M.K. (1985) On the evolution of parthenogenesis. II. Inbreeding and the cost of meiosis. *Evolution*, **39**, 1194–206.

Uyenoyama, M.K. (1988) On the evolution of genetic incompatibility systems: incompatibility as a mechanism for the regulation of outcrossing distance. In: *The Evolution of Sex: An Examination of Current Ideas* (eds B. Levin and R. Michod). Sinauer Associates, Sunderland, Mass., pp. 212–32.

White, M.J.D. (1973) *Animal Cytology and Evolution*, 3rd edn. Cambridge University Press, London.

Williams, G.C. (1975) *Sex and Evolution*, Princeton University Press, Princeton, N.J.

Williams, G.C. (1980) Kin selection and the paradox of sexuality. In: *Sociobiology: Beyond Nature/Nurture? Reports, Definitions and Debate* (eds G.W. Barlow and J. Silverman), Westview Press, Boulder, pp. 371–84.

Woese, C.R. (1983) The primary lines of descent and the universal ancestor. In: *Evolution from Molecules to Men* (ed. D.S. Bendall), Cambridge University Press, Cambridge, pp. 209–33.

CHAPTER TWO

Sex-specific reproductive patterns in some terrestrial isopods

K. Eduard Linsenmair

2.1 INTRODUCTION

Among crustaceans, only the order Isopoda produced many truly terrestrial species. Their successful invasion of the land required numerous preadaptations (Edney 1954, 1968; Warburg, 1968). The evolution of a direct ontogenesis without a planktonic stage was of special importance. A key development in this process was that of the brood pouch, formed by the oostegites, leaf-like coxal appendices.

In its 'terrestrial type', found in all higher Oniscidea (Hoese, 1984), this marsupium is not just a simple container, but protects eggs against desiccation and microbes and allows females to remove their brood from dangerous places and to carry them into favourable zones (e.g. thermal optima for embryogenesis). In ovoviviparity, developed for example by some reptiles (Shine and Bull, 1979), the matter which flows between mother and brood is restricted to oxygen, water and carbon dioxide: the eggs otherwise possess complete autarky. As has already been presumed by Patanè (1940), Akahira (1956) and others, the provision for developing embryos within the marsupium of higher oniscids apparently goes beyond this (Janssen and Hoese, in press). Therefore, 'the terrestrial type of marsupium may be regarded as an extension of the body cavity or a kind of uterus' (Hoese, 1984).

By producing yolk-rich eggs, brood pouches, a nutritive marsupial fluid, and by spending weeks in developing oocytes and carrying brood (Warburg *et al.*, 1984), females make high maternal investments. Isopod males, however, seem not to incur comparable paternal expenditures. Important consequences for sex-specific mating and reproductive tactics should result from these differences, according to sociobiological theory based on the hypotheses of Williams (1966) and Trivers (1972). Isopods and mammals show evident convergences with regard to basic features of reproductive behaviour. As crustaceans,

19

isopods are distinct from insects and arachnids in numerous traits. They are a diverse group, with many species adapted to greatly differing ecological conditions. These points, together with others, render isopods extremely interesting study subjects for the behavioural ecologist. They deserve far more attention than has been hitherto paid to them, especially with regard to comparative aspects, e.g. when examining general rules which can only be put to a critical test in systems which evolved convergently. This is valid for Crustacea in general: 'they show a greater diversity in social structures and mating systems than most other animal phyla' but they are very often completely 'missing from recent surveys of animal social systems and mating strategies' (Wickler and Seibt, 1981).

For years, our isopod studies were concentrated on the desert isopod *Hemilepistus reaumuri* with its highly evolved social behaviour and unusual ecology (Linsenmair and Linsenmair, 1971; Linsenmair, 1972, 1979, 1984, 1985, 1987). Later on, other subsocial species were included (Linsenmair, 1979, 1984, and unpublished), and it was only relatively recently that we started to investigate many non-social species. Since most of the studies on these non-social species, which represent more than 95% of all oniscideans, are still in their infancy, the picture that can be drawn on sex-specific reproductive patterns in terrestrial isopods is still very fragmentary.

An important part of the observations and data given below for *Porcellio linsenmairi* (nov. spec. Schmalfuss, in preparation) originate from an unpublished diploma thesis (Würsching, 1987), and some of the results relating to *P. scaber* are from another unpublished thesis (Golla, 1984).

2.2 ISOPODS AND METHODS

The isopods used in this present study originated mainly from different North African habitats. Species marked with * have also been, more or less extensively studied in the field. *Porcellio albomarginatus, P. hoffmannseggi, P. linsenmairi, P.spec.* (probably *obsoletus*) and three further, still undetermined, *Porcellio* species, live in semi-xeric to temporally xeric habitats, with *P. hoffmannseggi, P. linsenmairi* and two of the not yet determined species showing broodcare. These species are, according to our very preliminary observations, night-active, spending days in natural crevices, mostly under stones. Three *Porcellio spp.**, dwelling in different parts of the 'Island of Fuerteventura can also be categorized as semi-xeric species. These subsocial species are still undescribed (Hoese, in preparation). The species, mostly referred to on the following pages (and labelled *P. spec. I* i.e. an unnamed species) is a very conspicuous, large, highly colour-polymorphic, mainly day-active isopod distributed over a small area of the Jandia peninsula. All three species seem to be closely related to each other and to the North African *P. albinus.* All live on sandy soil and dig burrows, their reproductive cycles are apparently not tightly bound to certain seasons, but depend to a great extent on the very unpredictable rainfall. Three of the investigated species thrive under xeric conditions: *P. albinus*★, *P. olivieri*★, and *Hemilepistus reaumuri*★. For comparative studies, the Middle European mesic *P. scaber*★, *Oniscus asellus* and *Armadillidium vulgare* were observed. All these species were maintained and bred in the laboratory, where many different maintenance methods, which cannot be described here in any detail, were used.

The relevant behavioural reactions were either directly observed or recorded on videotape with an infrared sensitive camera (National WV–361 N/G, recorder: National Time Lapse NV 8030). In determining live weight of isopods, a Mettler AE 160 balance was generally used, with a Mettler ME 22 microbalance for smaller weights. For sterilization, males were irradiated with 150 Gy, using an X-ray instrument produced by C.H.F. Müller (Hamburg). In utilizing enzyme polymorphism for paternity determinations, the usual starch gel electrophoresis methods (Harris and Hopkinson, 1976; Blaich, 1978) were applied and adjusted to our needs (Körner, 1986). The polymorphic enzymes concerned were cytoplasmic malate dehydrogenase and phosphoglucose isomerase.

2.3 RESULTS

2.3.1 Mating tactics of males

In isopod males, we should expect a polygynous or promiscuous (Wittenberger, 1979) mating pattern. This assumption, however, will only hold true if the expenses a male has to meet to fertilize a single clutch remain on average considerably lower than a mean expenditure of a female on a single brood. Quantitative measurements of all relevant components of reproductive expenditures are not available for any isopod. Some of these costs cannot be judged from laboratory measurements, e.g. males' search costs.

(a) Copulation costs versus expenditures for clutch production

By relating dry weight of broods to residual weight ('somatic' weight: Sutton *et al.*, 1984) of females, a standing crop measure of their direct productive effort can be obtained. The brood pouch was to be included as well, since it is produced anew with every clutch. In males, one can proceed analogously by using dry weight of testes (including accessory glands). This crude measure can be somewhat refined by determining the corresponding calorific values, especially in isopods which have a high ash content in some of their body compartments. Investments measured in this way undoubtedly only mirror part of the costs (Hirschfield and Tinkle, 1975; Tinkle and Hadley, 1975). As a preliminary basis for revealing intersexual and interspecies differences, however, such measurements seem justified. At least they are preferable to purely theoretical assumptions which often led to an underestimation of ejaculate costs (Dewsbury, 1982; Nakatsuru and Kramer, 1982).

Tables 2.1 and 2.2 summarize data from three North African species, the subsocial *Porcellio linsenmairi*, living under semi-xeric conditions, and two species dwelling in xeric habitats: *P. olivieri* with no additional care for its progeny after delivery and the desert isopod *Hemilepistus reaumuri* with elaborate social behaviour. Relative clutch and brood pouch dry weights together (taken as a percentage of somatic dry weight of the females) exceed those of relative male gonad weights by factors of 7 to 21. (Values for *P. olivieri* can only be compared with those of the other species with reservation, since all the *P. olivieri* females used were not yet half-grown and were in their first breeding cycle. Preliminary data point to considerably higher investment in succeeding broods.)

The investments of all three species are notably lower than the figures given by Sutton *et al.* (1984) for seven Middle European species, in which weights of single batches of marsupial mancas averaged 24 to 47% of females' somatic dry weights. This pertains, in particular, to *Hemilepistus*, which expends on its sole brood a mean of only 7% (range: <2–10) of its somatic weight. The proximate means by which investment is lowered in the desert isopod is not a reduction in egg numbers since, with an average of about 80 (maximum 150–160) eggs per batch ($n=>>1000$), *Hemilepistus* falls into the upper range of the known oniscideans (Warburg *et al.*, 1984). The low investment is achieved by a reduction in egg size in relation to female weight (see Figure 2.1). The reproductive effort saved in direct investment could contribute decisively to making the extended brood care in this species feasible.

In males, too, there are evident, but much less spectacular species–specific differences (Table 2.2). These three species, however, provide no basis for an interspecific comparison from which any well-founded conclusion can be drawn as to which parameter of mating tactics is related to relative gonad size. Sperm competition has to be assumed to play an important role in all species (see p. 28), but its extent will certainly differ greatly from species to species, and this may be the decisive variable. In assuming – to be on the safe side – a ten per cent transfer of the entire dry mass of the gonads in each complete copulation (isopods possess two fully separated genital tracts; an insemination of all eggs, therefore, requires two 'single-sided' copulations, which will be termed here 'complete' copulations), males of *P. olivieri* could theoretically perform about 30, *P. linsenmairi* 105, and *H. reaumuri* 225 complete matings before their females' direct reproductive investments were balanced, provided that the costs per unit of dry weight amount to a similar sum in both sexes.

22

Table 2.1

Relation of dry weight and energy content of single broods (eggs/embryos) and brood pouches (oostegites) to fresh and residual (somatic) dry weight of females[1]

A Species (n) (average number of eggs/embryos)	B Fresh weight (mg)	C Somatic dry weight (mg) (joule/mg)	D Eggs/embryos dry weight (mg) (joule/mg)	E D given as % weight of C / as % energy content of C	F Brood pouch dry weight (mg) (joule/mg)	G F given as % of C / as % energy content of C	H Single egg/embryo given as % of C
Hemilepistus reaumuri (n=79) (87.1±28.6)	346.3 ±41.8	81.84±13.93 — 10.1	5.64±1.95 — 27.6	6.89±2.38 — 18.89	3.1 ±1.23 — 16.8	3.79±1.5 — 6.3	0.079
Porcellio olivieri[2] (n=22) (25.1±11.1)	72.6 ±16.8	16.2 ±5.63 — 10.5	1.38±0.48 — 24.7[3]	8.52±2.7 — 20.48	0.35±0.17 — 15.1[3]	2.16±1.05 — 3.1	0.34
Porcellio linsenmairi (n=15) (61.4±16.1)	143.17±30.23	35.93± 9.66 — 12.0	4.06±1.35 — 27.6[3]	11.3 ± 3.76 — 26.8	0.65±0.15 — 26.5[3]	1.81±0.42 — 4.1	0.19

[1] Most values are given as means ± standard deviation.
[2] Not yet half-grown individuals, see text.
[3] Maximal error 10%, in all other cases <1%.

Table 2.2
Relation of dry weight and energy content of gonads and accessory glands to fresh and somatic dry weight in males

A *Species* *(n)*	B *Fresh* *weight* *(mg)*	C *Somatic* *dry* *weight* *(mg)* *(joule/mg)*	D *Gonads* *dry* *weight* *(mg)* *(joule/mg)*	E *D given:* *as % weight of C* *as % energy* *content of C*
H. reaumuri (*n* = 25)	396.5±44.5	116.3±15.3	0.39±0.09	0.34
		10.6	23.2	0.74
P. olivieri (*n* = 16)	120.1±18.4	32.4± 5.8	0.61±0.14	1.9
		11.5	27.7 [1]	4.54
P. linsenmairi (*n* = 11)	161.7±31	45.4 ±■	0.45±0.14	0.99
		11.4	28.6 [1]	2.48

[1] Maximal error 10%, in somatic dry weight measurements <1%.

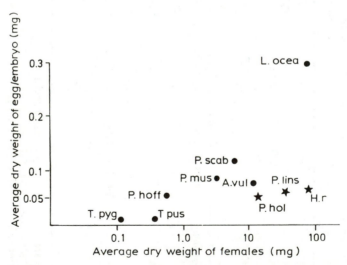

Figure 2.1 The relationship between average dry weight of females, from different oniscidean species, and average dry weight of single eggs or embryos. *: own data; remaining figures from Sutton *et al.*, 1984. (T. pyg) *Trichoniscus pygmaeus*, (T. pus) *T. pusillus*, (P. hoff) *Plathyarthrus hoffmannseggi*, (P. mus) *Philoscia muscorum*, (P. scab) *Porcellio scaber*, (A. vul) *Armadillidium vulgare*, (L. ocea) *Ligia oceanica*, (P. ol) *Porcellio oliveri*, (P. lins) *P. linsenmairi*, (H. r) *Hemilepistus reaumuri*.

In the monogamous *Hemilepistus*, males are usually permitted from two to six long complete copulations (5– > 15 min duration) by their female partner and a larger number of short ones (10– > 40). These last for between one and two minutes, or sometimes less than one, are usually single-sided and are always terminated by the female. In *Hemilepistus*, copulations require the preceding establishment of a social relationship between the partners of a pair, e.g. they have to be familiar with one another for a minimal time. This is highly dependent on the stage of the female's reproductive cycle and the prevailing climate (mean temperature), and ranges from about 12 h to many days. Females determine the temporal distribution of all copulations within the respective phase after pair formation, and the course of each copulation. Long copulations are usually only permitted approximately after a third to half of the time span between pair formation and parturial moult has elapsed. This time period extends from 3–7 weeks in the field, and is shortened to 10–15 days when temperatures reach 24–28°C. These peculiarities render it impossible to investigate the upper physiological border of the male's mating capacity. When males (at temperatures of 24–28°C) were forcibly separated from their previous female and, every 3–4 days, given a new virgin female in a relatively advanced state of its reproductive cycle, all six test males succeeded in fully inseminating each of the six to ten females ($n = 31$ females surviving until delivery).

We know nothing about the copulatory behaviour of the two remaining species under natural conditions. In the laboratory, their copulation frequency is higher than that of *Hemilepistus* (15 and more complete copulations per 24 h over periods of many days and presumably weeks). Copulations of these and other promiscuous species were always of notably shorter duration than in *Hemilepistus* (in *P. linsenmairi*, for instance, always less than two minutes. This certainly does not tell anything about the amount of ejaculate size in an interspecific comparison, but may point to effects of differing selection pressures caused by intrasexual competition in males. In *Hemilepistus* and the transient pair-forming *Porcellio* species, owing to site and time of copulations, it is extremely improbable, and has never been observed, that another male will interfere while a pair copulates. Such interference, however, is to be expected, being very common in all aggregating species.

(b) Female availability and intrasexual competition in polygynous males

As preliminary experiments with the non-social *Porcellio olivieri* and *P. albomarginatus* and *P. spec.* (*obsoletus?*) have demonstrated, males kept in the laboratory without competitors may achieve much higher successes than *Hemilepistus*. Since hardly any pre-copulatory time has to be spent in these species, males succeeded in fertilizing dozens of females within the time required for a female to produce a single brood. To obtain the maximum gain from their high mating capacity, (a) receptive females should be available during the entire reproductive phase, (b) these receptive females should be found quickly, and (c) males should be able to monopolize these females by effectively excluding competitors for prolonged periods.

Numerous isopods exhibit very high local abundances (Sutton, 1968; Warburg *et al.*, 1984), thus high encounter rates should be achievable in several species. However, owing to the fact that (a) females produce a very small number of broods per productive cycle (in most known species only one to two: Warburg *et al.*, 1984; Warburg 1987), (b) they are unreceptive for prolonged periods, e.g. generally while carrying brood in their marsupium, and (c) they are strongly dependent on the annual climate cycle and, therefore, rather well synchronized, the first precondition is certainly not fulfilled. Receptive females are, at least in temperate and subtropical regions, very heterogeneously distributed in time, reducing males' potential reproductive success. Figure 2.2 gives an example. In *P. scaber*, a large number of receptive females are present only during the beginning of the annual reproductive phase and eventually for a short period before the second parturial moult when, depending on climate, a second brood is produced. At other stages, up to 100% of all females exceeding a body length of 10 cm carry brood and are unreceptive. Only those small females (with low fecundity) which reach a length of 6–8 mm early in summer, still become partly receptive during the same year. Males, however, retain their mating readiness, irrespective of female availability. Therefore, the operational sex ratio (Emlen and Oring, 1977) should be extremely male-biased throughout most of the reproductive phase. Strong inter-male competition is to be expected. What is its outcome?

Four to seven males of *P. scaber* and of *P. linsenmairi* were kept together in small containers (10 × 10 cm). When receptive females were placed in these containers, mating successes were very unequally distributed among males. Aggressive encounters led to the formation of rank orders. In all containers, it was the two heaviest males that obtained most matings (44 from 49 complete copulations), although differences in weight between members of the same group did not exceed 10% in many cases. Fight duration was not more than a few minutes and led to mostly linear dominance relationships which remained stable for many days. Low rank positions did not prevent males from attempting to mate.

(c) Female guarding

In several aquatic crustaceans, especially in the Peracarida, female receptivity and parturial moult are tightly coupled, and fertile copulations are restricted to a short period during and/or after ecdysis. In these species, precopulatory guarding behaviour, a 'mate guarding monogamy' (Wickler and Seibt, 1981) is widespread. Apart from the Tylidae and a few others, this interdependence

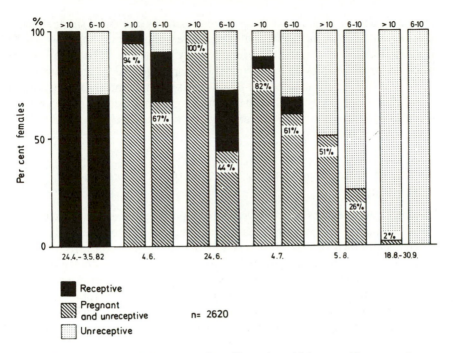

Figure 2.2 Percentage of females in *Porcellio scaber* which were either receptive, or unreceptive – without carrying brood, or pregnant and then unreceptive, too, at different times of the reproductive cycle. Samples were taken from the field at the indicated dates. 6 – 10 and >10: females with body length between 6 to 10 mm or females exceeding 10 mm. Each sample contained about 300 to 500 females.

between receptivity and moult was abandoned in the evolution of oniscideans, and the phase of receptivity became greatly prolonged (Mead, 1976).

In all gregarious isopods which we have observed to date, males follow a promiscuous mating pattern, trying to mate with every receptive female they encounter. Clear hints as to male choice are only observed rarely, when males come into contact with two females simultaneously. We could never find indications of long lasting guarding behaviour in these species. Guarding an individual female in a large aggregation could only be realized through permanent physical contact, as is observable in, for example, the precopula of *Asellus* (Ridley and Thompson, 1979). In the higher oniscideans, a comparable mate guarding strategy has not been developed, most probably because it is incompatible with the requirements of their terrestrial life, and not because it is always less profitable than a promiscuous mating tactic.

In males of the non-aggregating, burrow-digging Canarian and North African *Porcellio* species, mate guarding is common (Linsenmair, 1984, and unpublished). In at least one species (*P. spec. I*) males follow different tactics depending on their size. Large males form temporary pairs with single females (see section 2.3.2), while small males search around, attempting to copulate more or less forcibly with every female they encounter. The time cost of guarding single females obviously pays off for large males, since otherwise they could have easily adopted the tactics used by small males. Monogamy in *Hemilepistus* probably evolved from such temporary mate guarding behaviour, as is found in these transient pair-forming *Porcellio* species (Linsenmair, 1979, 1984).

(d) Is sperm quantity more important than temporal order of copulations?

Do males guard females in order to secure a special copulation or to secure as many copulations as possible, because it is the quantity of sperm deposited which decides their success? The very few data in the literature from promiscuous species in which allozyme polymorphism was used for paternity determination of field-collected females, do not point to a sperm precedence according to temporal order of copulations (Sassaman, 1978; Johnson, 1982). This is contrary to most insects investigated to date (Parker, 1970; Gwynne, 1984). Sperm of isopods (which are immobile and packed in bundles wrapped in a membrane (Fain-Maurel, 1970; Cotelli *et al.*, 1976)) from different copulations, however, appear to become mixed before fertilization occurs.

This question was followed up in *Hemilepistus* in two ways. First, by pairing females with X-ray irradiated males and, thereafter, with normal males, or vice versa, and determining the number of developed embryos and undeveloped eggs shortly before delivery. Second, by pairing females with males differing in their allozyme patterns and carrying out electrophoretic paternity determinations when the young were half-grown. The problems which may result from using irradiated males have been amply discussed (Gwynne, 1984). Sperm mobility, which could be negatively influenced by irradiation, plays no role in isopods. With regard to the general behaviour of males, negative effects were never detected during the critical time-span of the copulation phase, if the male irradiation dose did not exceed 150 Gy. Above all, results obtained by using allozyme polymorphism, on the one hand, and those gained by utilizing X-ray treated males, on the other, showed no significant differences. Therefore, these data can be taken into consideration without reservation.

The starting point of these investigations was the finding that sexually active *Hemilepistus* males did not differentiate between females at differing stages of their reproductive cycle, apart from those performing or having performed parturial moult. The fact that virgin females were never preferred over those which had already been paired for the majority of their mating phase and had lost readiness for further copulation with their first male, would easily be understandable within the framework of a fertilization modus with last sperm being used first. But, as the results summarized in Figure 2.3 and in Table 2.3 demonstrate, this is definitely not so. On the contrary, it is one of the most obvious and highly significant results that males which

Table 2.3

Reproductive successes of X-ray irradiated and normal males respectively according to time of consorting during the phase from pair formation until parturial moult

A n (Broods)	B Average time span from pair formation until moult (days)	C Relative separation time	D % fertilized eggs: first male (number of eggs/ embryos)	E % fertilized eggs: second male (number of eggs/ embryos)
				Significance D : E
21	22.9±3.7	Beginning of second half corresponding to day 12–14 of B	57.5 ($n = 1381$)	42.5 ($n = 1022$)
				$p < 0.0001$
31	22.9±3.7	Beginning of second third corresponding to day 15–18	80.4 ($n = 2629$)	19.6 ($n = 640$)
				$p \ll 0.0001$
72	12.6±1.3	Second half corresponding to day 6–9 of B	84.4 ($n = 6439$)	15.6 ($n = 1189$)
				$p \ll 0.0001$
7	12.6±1.3	Last fifth corresponding to day 10–11	100[1] ($n = 648$)	0[1] ($n = 14$)

			Controls	
	Males	% fertilized eggs	% eggs non-fertilized	Average clutch size
15	normal	94.8 ($n = 1433$)	5.2 ($n = 74$)	100.5±13.7
19	irradiated 150 Gy	0	100 ($n = 1924$)	101.3±12.6

[1] The 14 eggs, which could formally be attributed to the second male, were unfertilized eggs in clutches fathered by normal males. Since in those clutches a rate of 5.2% (see controls) of non-developing eggs is contained, it is assumed that these 14 eggs fall into this category. In four of seven cases the normal male was second with no single egg having been fertilized.

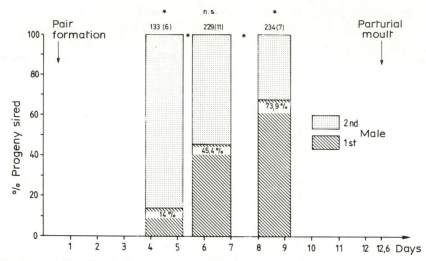

Figure 2.3 Fertilization successes of first and second males, exchanged at different times within the mating phase. Males differed in allocymes; paternity was determined when offspring was half-grown. Numbers above columns (without brackets) indicate number of young tested. Numbers with brackets give number of broods. *above and between columns: null hypothesis of equal distribution of reproductive successes within and between columns, respectively, has to be rejected ($p < 0.0001$, chi-square tests).

replace their predecessors only late in the pair phase have little or no paternity chances. As video recording of the behaviour of such pairs has clearly demonstrated, reproductive failure of late males is not caused by females preventing them from copulating. Figure 2.4 shows one common picture when pairs are maintained under high temperatures. Figures 2.5 (a) and (b) demonstrate what happens when the first male is removed and the female is given a new male 16–24 h later. The second male is permitted long copulations, mostly within 24 h of pair formation, and yet, from the point of view of offspring sired, he often 'goes out empty-handed'. Since these copulations did not differ in any respect from those occurring earlier, it has to be assumed that large quantities of sperm are transmitted, but they are evidently prevented from being used for fertilization.

If the first male is replaced shortly before 'half time' of the pair phase, the second gains a definite advantage in terms of sired progeny. Only in the group of those males exchanged almost exactly in the middle of the pair phase, are the first and second males, on an average, equally successful. When each of the 149 broods is examined separately, it is striking how small the actual percentage of clutches is in which the null hypothesis of a uniform distribution of reproductive success in both males cannot be rejected, namely 14.8% ($n = 22$). (In nine of these 22 cases it was probably small progeny number alone and not equality which led to non-significant differences!)

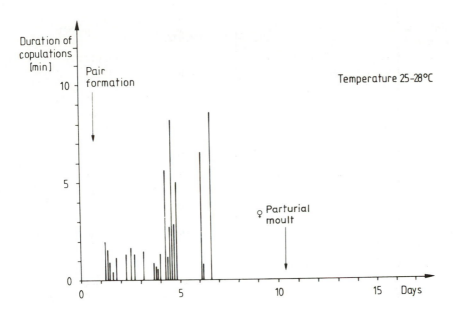

Figure 2.4 Length of copulations (in minutes) and their temporal distribution (in days). Due to high temperatures used in the laboratory experiments, the time span between pair formation and parturial moult is, when compared with natural conditions, considerably shortened. Under lower temperatures no copulations take place during the first days (up to more than one week), otherwise no fundamental differences in dependence on temperature were found.

(e) Reproductive tactics of males: discussion and conclusions

Sex-specific differences in mating patterns and reproductive strategies of most known isopod species tend to fit the general picture. Females invest heavily in terms of energy and time in their progeny, while males, apart from *Hemilepistus*, avoid costs in the currency of parental effort and try to maximize their reproductive success by investing all their expendable resources in mating effort. The high number of copulations observed, especially in *P. linsenmairi*, *P. olivieri* and in other promiscuous species, which we have begun to study, prove that single copulations are certainly not very expensive, most probably much cheaper than assumed above. Cheap copulations, limited availability of females during extended periods of the reproductive cycle and strongly male-biased operational sex ratios must lead to strong male competition with resulting rank orders at least in some species.

Are dominant males, in those species forming aggregations, able to monopolize many females, thus gaining very high reproductive benefits? Theoretically, at least, conditions seem favourable. But, in reality, constraints should set rather narrow limits on the maximum attainable success. These

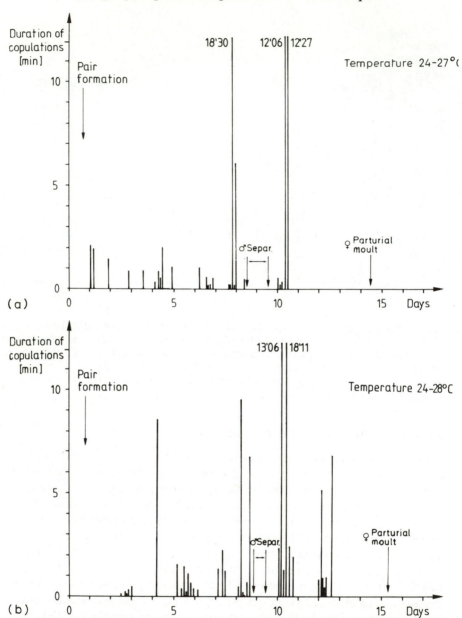

Figure 2.5 Copulation behaviour of females, losing their first male and pairing anew with a second male. (a) In this female additional long copulations would probably not have taken place without loss of the first male. (b) This female had, when losing its male, not yet terminated its copulations phase. In (b), the second male had still rather good paternity chances, whereas the second male in (a) most probably became a pure stepfather, despite its long copulations.

constraints are deficiencies in the realm of sensory physiology. In the oniscideans, sense organs for any form of remote sensing of conspecifics are absent. Therefore, isopods are only able to control that part of their immediate vicinity which they can reach with the tips of their second antennae. Outside this range they are even unable to recognize a conspecific as such, let alone to categorize it. Thus, their capability in detecting receptive females in due time before they have been discovered by subordinate competitors must be severely restricted. It is impossible for these males to simultaneously guard many mobile females inside and outside their retreats for weeks and to reliably fend off competitors during this long period. Suitable preconditions for a resource defence polygyny (Emlen and Oring, 1977; Wittenberger, 1979) cannot therefore be exploited by isopod males.

Under experimental conditions, it can easily be shown that, with a single complete copulation, not only is fertilization of the first clutch secured, but the transferred sperm quantity is, in a number of species (e.g. *P. linsenmairi* and *P. scaber*) (Vandel, 1937, 1941; Lueken, 1962; Johnson, 1982), sufficient (and can be stored long enough) to fertilize the life-time egg production of the female. In most species, however, females permit more than one, and apparently often far more than one, copulation. If several to many complete copulations are needed — on the male's side – to fertilize a clutch successfully, cost of ejaculates is certainly not meaningless, and this could result in another constraint being imposed on the reproductive success of dominant males in competitive situations. Under favourable conditions allowing for high copulation frequencies, a reduced number of sperm-bundles could become a limiting factor for the reproductive success of promiscuous males engaged in an indirect competition via sperm quantities.

Where many copulations are the rule, it has to be assumed that the average single copulation is of only limited reproductive value for the male. If there is a clear sperm precedence with last sperm being mainly used for fertilization, males should, when given a choice, preferentially mate with females shortly before or during parturial moult. This is found among oniscideans in the mostly semiterrestrial Tylidae (Mead, 1967) and also in some species of the higher Oniscidea (Lueken, 1968; Mead 1976). To secure this copulation, the evolution of a postcopulatory or pre and postcopulatory mate guarding behaviour is conceivable, as is realized in some Tylidae (Mead, 1967). If no sperm clumping exists and sperm from multiple copulations are mixed and quantity is the decisive factor as to success or failure, it would pay males to also develop a mate guarding behaviour. The two cases should be distinguishable according to the temporal distribution and frequency of copulations, as intended by the male and according to the type of guarding, whether pre or postcopulatory.

In *Hemilepistus* as well as in the Canarian *Porcellio spec. I*, where we have extensive data on the temporal distribution and frequency of copulations and on the time courses of male and female mating behaviour, all results point to the second type of guarding. Males copulate as soon and as often as they are allowed by their females and we were unable to find a preference for females which were near to their parturial moult. The guarding behaviour of males cannot be described as being either pre or postcopulatory, but is both. In all primary pairings – in *Hemilepistus* and the pair-forming *Porcellio* species – one to several (in extreme cases up to 30) days always pass before first copulations are allowed by females, given temperatures corresponding to those prevailing under natural conditions during this time. If pairs stay together undisturbed, females lose their readiness to copulate well in advance of the moment when – in the *Porcellio* species – they are left by their males or become unattractive to other males. Thus, guarding behaviour seems to serve the monopolization of females, on the one hand, and the protection of paternity, on the other.

In the majority of organisms with an internal fertilization and with multiple inseminations, investigations have proved that the last male has advantages (Parker, 1970; Gwynne, 1984), although the processes leading to sperm displacement are generally unknown. In some other species, first copulations are the most successful (Walker, 1980). In *Hemilepistus*, and probably also in the pair-forming *Porcellio* species, with their temporally extended mating behaviour consisting of more than 30 complete copulations during a single mating phase in some pairs, the situation is different. Neither last nor first copulations are reproductively most important ones as long as the usual copulation phase could pass undisturbed. Females in these species obviously have a rather large sperm storage capacity, only a minor portion of which will be used. Since all results point to sperm mixing, chance will decide which sperm bundles are used for fertilization. The prospects for males, therefore, will improve with the number of sperm they are able to throw into the game. The capacity of the female's store, however, is definitely limited. Females, on the other hand, do not inform males when this point is reached. In *Hemilepistus* they accept additional ejaculates, but the sperm transmitted in late copulations are either removed or brought into a spatial position or physiological condition in which fertilization chances become minimal. What is decisive in these isopods is to fill the store, and this requires more than one copulation (or unusually long copulations as observed in *Hemilepistus*) during a certain time span.

The fact that in the sperm competition trials, both males had similar successes in only a very limited number of cases suggests that the filling up of the sperm reservoir does not take place in small steps which are equally distributed throughout the middle period of the pair phase. Only a few semi-long copulations following each other in rapid sequence (i.e. within a few hours) or a single very long copulation around half-time will be decisive for the male's final reproductive success. This assumption is supported by video recordings of copulation patterns, showing that first copulations, lasting five

34

minutes or more, usually coincide with this critical time span. Under our laboratory conditions (22–28°C) these copulations are usually performed at night. Since experimental separation of pairs was carried out during the daytime, these important copulations either had or had not already taken place. The interesting question, as to why *Hemilepistus* males never voluntarily leave their females and attempt a successive polygyny after having performed the long copulations, can not be followed up here (Linsenmair, 1979, 1984, and in preparation).

Males are – as stressed above – unable to assess females according to their reproductive prospects. Frequent problems with pair cohesion, i.e. males expelling females from their common burrow during or shortly after parturial moult, were only observed in pairs formed during the last 48 h before the females' parturial moult. Apart from the fact that they would have lost less time by avoiding such a female from the very first, the males behaviour does, by no means, guarantee their not being totally or at least very severely cuckolded. Females may have filled up the reproductively interesting part of their sperm reservoir much earlier, as our above results convincingly demonstrate. Under natural conditions, local inbreeding is very effectively avoided in *Hemilepistus*, so a late second male will not even be distantly related to his foster young. Here, as in some other cases, males are manipulated to their utmost disadvantage but to the great advantage of the manipulating female (Linsenmair, in preparation).

Hemilepistus (and to a lesser extent the transient pair-forming *Porcellio* species) deviates, in many respects, from the usual oniscidean behaviour. This may also apply to its mating behaviour, which could have been fundamentally reshaped according to new selection pressures arising during the evolution of the highly specialized social behaviour. Therefore, data gained in other species, which form only transient or no pairs at all, are needed, before generalized assumptions can be formulated. Our efforts to answer the questions as to what extent temporal order of copulations and their frequency determine reproductive success in the promiscuous *P. linsenmairi* have failed to date, because sperm of the phospho glucose-isomerase-polymorphic male used were not equivalent when competing with each other. Independent of the temporal order and different copulation frequencies, one genotype was always overrepresented. There was, however, no reduction in rate of fertilization in controls where females were inseminated by males of the inferior genotype only.

2.3.2 Reproductive tactics of females

In describing the males' mating patterns, some traits of females' mating and reproductive tactics have already been mentioned. Only two additional topics will be raised here before discussing female reproductive strategy in general: male choice and brood care in females.

(a) Female choice in isopods?

We have seen that (a) strong inter-male competition is to be expected in many species and that (b) males can hardly succeed in monopolizing females if such behaviour runs counter to the interest of the females. These conditions should favour female choice behaviour. Since females possess only a very limited reproductive capacity and must have great difficulties in compensating for offspring losses by producing larger or additional broods, they should be selected to avoid all risks of reducing the survival probability of any of their progeny, including that of choosing a male of low genetic quality (Maynard Smith, 1987).

Do females choose certain males and reject others? This is unequivocally so in the Canarian *Porcellio spec. I.* Here, receptive females search for large males which dig special, deep copulation burrows (usually stocked with ample food by the males). According to field and laboratory observations, females join these males for many days, up to three weeks. Under laboratory conditions females behave as follows. During the first week, females resist male copulation attempts, then permit many copulations for the next four to seven days, after which they again become unwilling to copulate, but still remain attractive to their males (and other males as well) for a further week. Females were observed to copulate only within the burrow. Thus, males which were not expelled from their copulation burrow by a superior competitor, were guaranteed absolute copulation privileges. All females resisted the forcible copulation attempts of small males, large females with special vehemence and always with success, while small females sometimes gave in. Small males were not observed to dig copulation burrows.

In observing *P. linsenmairi* females which are consecutively placed with different males, one gets the impression of very unselective mating behaviour. These females always copulate with every male, mostly within a few minutes following their first encounter. In detailed observations, however, behavioural differences come to light. In encounters with large males (between 115–80% of the female's weight) females permitted copulations, on the average, within the first minute of contact (32–78 s, $n = 18$ copulations). With small males (80–60% female weight), in contrast, this time span always exceeded three minutes in $n = 23$ copulations. These delays are caused by the females not standing still, but withdrawing, with males attempting to follow females by maintaining antennal contact. Singly kept, small males never suffered any reproductive losses through the females' behaviour. When females were placed in small containers with two males of different sizes, however, it became apparent that the ratio of 30:4 copulations obtained by the larger males was attributable, to a large extent, to the females' behaviour *vis-à-vis* small males. Females did not only delay copulations by their frequent withdrawals, resulting in larger males often becoming attentive and having the

36

opportunity to interfere, but their discriminative behaviour went one step further. Those females, which had already had contact with both males, sometimes attacked small males violently, but allowed the large males to copulate without resistance.

(b) Mate choice in *Hemilepistus*

In the monogamous *Hemilepistus*, those individuals digging a new burrow or gaining possession of an existing den can choose amongst potential mates by indirect means, the local partner market being effectively exploited. The burrow owner allows admittance to a conspecific of the opposite sex only after a prolonged ritual (Linsenmair and Linsenmair, 1971; Linsenmair, 1985). If there is more than one applicant in the catchment range of the burrow, these competitors will engage in vigorous fights, with usually the largest (heaviest) applicant being the winner (Figure 2.6). The perseverence time (from first contact at the burrow entrance until admittance of a prospective pair partner) required by the burrow owner is only slightly influenced by the attributes of the applicant; it does not change, for example, when many fights lead to multiple replacements of the applicants, but is almost entirely determined by the reproductive physiological state of the burrow owner and by the prevailing temperature conditions (Linsenmair, 1984, and in preparation).

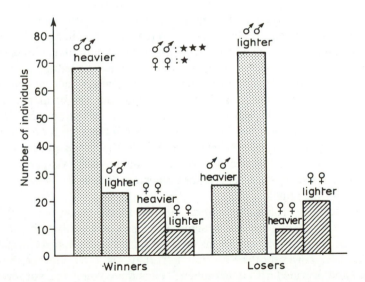

Figure 2.6 Outcome of aggressive encounters between competing males or females while striving for admission into a burrow, owned by a potential mate (in the field). Winners are, on the average, heavier than losers, ***($p \ll 0.0001$) *($p = 0.03$).

Mate choice in *Hemilepistus* not only helps the female to avoid 'bad genes', but also provides other direct benefits. Winners are usually larger than losers. Larger males are more suited for burrow defence, e.g. they can form better 'entrance plugs' against intra and interspecific competitors attempting to take over a burrow by force. The situation is, however, often reversed, with males being the burrow owner and females the applicants for burrow entry. A larger female may own better genes, which, in a strictly monogamous species, should be as interesting for a male as for a female. In addition, such a female produces, on the average, a large brood. It may thus pay a male to prolong the admittance procedure as well.

(c) Broodcare

As already stressed, the capacity for compensating reproductive losses via production of additional progeny must be severely limited in isopod females. Another way of gaining advantages in the intraspecific strife for maximal relative reproductive success and of coping with increasing reproductive losses under changing ecological conditions would be to invest even more in brood-care, which is carried on anyhow and, in this way, to increase the number of surviving young. This could be achieved by two different evolutionary routes.

First, the time spent by the hatched juveniles in the brood pouch could be gradually prolonged. Thus offspring would be released at later developmental stages, when they are less endangered, e.g. by adverse climatic conditions and/or predators. A reduction in number of progeny per brood seems an unavoidable prerequisite if this route is taken. The formico- and termito-philous Squamiferidae have obviously adopted this evolutionary direction. In the formicophilous *Plathyarthrus hoffmannseggi*, a female produces in each of its two to three broods not more than two to eight offspring which, at delivery, reach an average of 47% of the somatic dry weight of the female (Sutton *et al.*, 1984). This development has reached its extreme in *Exallon-iscus maschwitzi* (Ferrara *et al.*, in press), which lives in association with the ponerine 'army' ant *Leptogenys mutabilis*. Within the brood pouches of the females of this species, only two embryos are found at an early stage, of which one disappears and only one is developed reaching, at delivery, about half the body length of the female.

In the second evolutionary line, no prolonged brood carrying was developed. Young are born at the same developmental state with the same size (or even considerably smaller, as in *Hemilepistus*) as in non-social species. After delivery, however, they are not left alone by their mother, but are protected and provided with nourishment. Thus these young are not forced at the early stages to leave their protected, microclimatically favourable retreats to forage outside and then to seek shelter again. According to our present knowledge, this evolutionary route has been followed at least twice: first, in the ancestors of the genus *Hemilepistus* and second, within the genus *Porcellio* (and there probably more than once) (Linsenmair, 1979, 1984). With the exception of *Hemilepistus*, females alone provide all broodcare, as would be expected from the sex-specific differences in the distribution of reproductive effort in the non-broodcaring species.

Broodcare in isopods consists of the following: (a) in all species, building or taking possession of a suitable natural retreat offering a bearable microclimate to the very sensitive young; (b) in some species, cleaning this home of faeces and old food particles; (c) in all species, protecting the young against foreign (in some species only adult) conspecifics and against some potential enemies of the young or interspecific competitors interested in the den; and (d) above all and again in all species, providing the young with all necessary food for 2–6 weeks. In some species, this provisioning is continued for a few additional weeks, although the young at that time are already partly provisioning themselves. As far as we know, all species providing additional broodcare after delivery, with the exception of *Hemilepistus*, are multiparous and at least bivoltine. All *Porcellio* females leave their young long before these have reached sexual maturity. In *P. linsenmairi* females only recognize the site where their young remain by means of individually-specific chemical marks, which are produced by the female and which adhere to the ground. In *Porcellio albinus*, which has even invaded the sand seas of the central Sahara, and in the Canarian *P. spec. I* (and most probably in the other two Canarian *Porcellio* species, as well), not only is the burrow marked with an individually-specific chemical label, but young are also recognized family-specifically by means of variable discriminator patterns, as in *Hemilepistus* (Linsenmair, 1972, 1985, 1987). *P. linsenmairi*, on the other hand, accepts any young staying in its retreat, while the broodcaring *Porcellio albinus* and the Canarian *P. spec. I* expel alien young without seriously attacking them. In *Hemilepistus*, however, small alien young are fiercely attacked, where possible hunted, and, if caught, cannibalized (Linsenmair, 1987). These findings point to essential differences in population structure (viscous versus fluid) in the above-mentioned subsocial species.

(d) Reproductive tactics of females: discussion and conclusions

We expect females, which have a very limited reproductive capacity and invest heavily in their offspring, to choose amongst available males according to attributes which indicate the genetic quality of the possible prospective fathers of their brood (Bradbury and Anderson, 1987). The possibilities for oniscidean females to achieve this discrimination must be very limited, due to the fact that they are poorly equipped with efficient sensory organs. Our findings, in this respect, are that females prefer larger males either by direct choice (as in the Canarian) or by indirect means, by letting males fight for access (as in *P. linsenmairi, P. scaber, H. reaumuri*). In the pair-forming species, males accepted as social partners did not become sexual partners, under natural conditions, until a certain time period had elapsed, usually a few days. Males, eliminated during this time through competitors or by other reasons, including genetic ones, are thus prevented from fathering any progeny. Through her initial coyness, therefore, the female gains the opportunity for further indirect examination of the relative quality of the chosen male, compared with those still available, during a time span in which she can afford to remain unfertilized.

In oniscids, as in many other crustaceans, there is no definite number of moults. Growth continues after sexual maturity has been reached. Large males are relatively old males which have demonstrated their ability to survive for an extended time. Females choosing large males eliminate all those which die early due to genetic failures (Howard, 1978; Halliday, 1978), and those which remain small (due to genetic reasons with proximate causes such as metabolic deficiencies or heavy parasite loads: Hamilton and Zuk, 1982). Absolute choice criteria can only be used when (a) males of the preferred type are usually available and (b) the costs of searching for a male are low. Such conditions are characteristic of the populations of the Canarian *P. spec. I.* These isopods live in dense populations (depending on the substrate, 15–21 individuals per square metre) and show an unusual sex ratio, which will not be discussed here, of 1.47 males to one female (field data, $n = 1009$; laboratory data, gained by breeding, gave a ratio of 1.46, $= 794$). When only a limited number of males are present and when searching for additional males would be unprofitable, females should use relative criteria. Inducing males to fight and accepting the winner is probably the most effective and cheapest means of choosing the best male at a certain place and time.

As already stressed, a single complete copulation would suffice (e.g. in *Porcellio albinus, P. linsenmairi, P. scaber, P. hoffmannseggi* and all three Canarian Porcellio species) to fertilize a large proportion of a female's oocytes and often its entire life time production. Sperm not used for fertilizing eggs of a previous clutch can be stored for prolonged periods, this being well known for many years (Vandel, 1937, 1941; Lueken, 1962). Despite these circumstances, females of all the species we have observed to date (and on which data exist in the literature allowing inferences to be drawn as to the frequency of copulations) permit more than one copulation, often many more. Why? Females, at least the larger ones, can resist copulation attempts of males without great effort. They may practice this for long periods, e.g. in *Porcellio scaber* where females allow only a very limited number of copulations during a reproductive cycle, but remain fully attractive until parturial moult has terminated. The same holds true for all pair-forming species, where females always terminate the copulation phase, normally well in advance of their parturial moult and still remain as attractive to males as before. Therefore, the argument that the costs of avoiding copulations would be too high is certainly not applicable to isopod females, with the exception of very small individuals in some species.

Since all our data demonstrate convincingly that isopod females are in complete control with respect to their mating behaviour, it has to be assumed that females profit by allowing several to many copulations. In those species where pairs are formed and where females gain from the male's presence, multiple copulations could be a means of establishing a prolonged pair bond by the female. This could be achieved by 'withholding' all information regarding her reproductive state, especially about the most profitable time for reproductively successful copulations. This is comparable to the tactics of concealed ovulation discussed, for example, in the context of human monogamy (Alexander, 1979; Alexander and Noonan, 1979; Strassmann, 1981; Wickler and Seibt, 1981; Linsenmair, 1979, 1984). Such tactics however are doubtless derived, and not primary, behaviour. What could the advantages of multiple copulations be in polygynous and promiscuous species?

Many isopods live in more or less large aggregations, due to the island-like distribution of essential resources. Non-aggregating species are also, owing to special microclimatic and other ecological requirements, usually clumped in space and time. Under these conditions, a high degree of inbreeding seems unavoidable, unless effective mechanisms have evolved to prevent it, such as emigration or special forms of mate choice which guarantee outbreeding (Bateson, 1983). We know that *Hemilepistus* has developed such mechanisms (Linsenmair, in preparation), but have found no comparable methods of incest and local inbreeding avoidance in the aggregating promiscuous species. If local inbreeding is common, then multiple insemination could counter it to some extent, especially when coupled with some preference for encountered partners which do not originate from the female's own aggregation (Sassaman's 1978 results on *P. scaber* could point in this direction).

The assumptions concerning local inbreeding gain support from the fact that sex ratios in many species deviate from equality, without differing costs of male and female progeny production (Charnov, 1982). In most species with unequal sex ratios, females outnumber males, with proportions of males varying between 20–40%. In 21 samples (with a total of 2620 individuals) drawn from different populations at different times of the year, we found a mean of $31.6 \pm 6.4\%$ in *P. scaber*. This value conforms almost exactly to a sex ratio of one-half, calculated as the ideal proportion by Hamilton (1967, 1979) under conditions of local inbreeding.

As we have seen, two evolutionary ways are open for developing an extended broodcare. Whether the first or second route is taken in evolution may depend upon the spatial and temporal distribution of essential resources and the nature of the selection pressures acting against small young. Where a stationary life is prevented, as in *Exalloniscus*, living in association with a

41

nomadic ant and where, possibly, predators are the most important enemies of young, only the first way is open. If resources, however, allow a stationary life for the weeks or months of broodcaring, and if life of the very young is predominantly threatened by abiotic (climatic) factors, then the second way may be the better choice, since a prolonged brood-carrying may result in considerable disadvantages for the female.

There is no room to discuss the many peculiarities of *Hemilepistus* here in a general framework of oniscidean reproductive tactics. only one point seems to explain many of the specialities of *Hemilepistus*. *Hemilepistus reaumuri* and *Porcellio albinus* live in many areas side by side. They are of equal size, show at many sites a complete overlap in their food, differ only slightly in their daily activity patterns during part of the year, are very similar in many ecophysiological properties, and yet show fundamental differences in their life history patterns. While *Hemilepistus* is strictly monogamous, with extensive cooperation between pair partners, and semelparous, with a temporally very extended biparental broodcare, *P. albinus* forms only transient pairs, which dissolve before parturial moult, thus showing a successive polygamous mating pattern. Broodcare is performed by the female alone, with the period of broodcare being much shorter than in *Hemilepistus* (two compared to more than six months, given *Hemilepistus* parents survive that long). Furthermore, *P. albinus* is multiparous, at least bi- eventually trivoltine and may survive two years. Females born early in the year breed in the year of birth, and a number of additional differences in their behavioural and reproductive physiology could be enumerated.

In searching for the decisive ecological factor which could be responsible for the great differences in the evolutionary courses of the two species, one finds, in looking at the momentary situation (which of course may be misleading) only a single one, namely the substrate in which burrows are constructed. In *Hemilepistus* it is solid soil which becomes extremely hard when dry. Since *Hemilepistus* lacks effective digging adaptations, excavating a new burrow is a slow process which can only be started under mild climatic conditions, in spring. Later in the year, new burrows cannot be constructed. Thus, burrows are extremely valuable resources which can be secured against intra and interspecific competitors only by permanent guarding, a requirement which a single individual cannot meet. *P. albinus*, on the other hand, digs in light sandy soil, and can always reach a protective depth by digging for several hours during a single night. There is much less competition for these burrows. When, in *P. albinus*, the young are ready to care for themselves, their mother can leave, excavate a new burrow and produce a new brood. At a corresponding time, *Hemilepistus*, in contrast, could never dig a new burrow and, therefore, is not free to leave, although its young would have good prospects of surviving (given that the departing adults would not compete with them for burrows).

Summary

In order to secure a burrow in time, *Hemilepistus* females are forced to fit their reproductive cycle tightly into the course of the year, and this causes a rather strong synchronization among females, with the consequence that males' chances for successive polygyny are very reduced. In *P. albinus*, in contrast, females are less well synchronized, a second brood (at least) occurs and young females become partly receptive at a time when older ones are mostly pregnant. In a situation in which conditions change in such a way that they approach the present ecological situation of *Hemilepistus*, the only means for both sexes to increase their reproductive success on the adaptive level previously reached (i.e. on the basis of temporal pair formation and broodcare of females) is to stay together and to improve survival probability of their sole batch of young. Many of the peculiarities in the life history of *Hemilepistus* can be plausibly explained along these argumentative lines, although further elaboration is necessary.

2.4 SUMMARY

Sex-specific mating patterns and reproductive tactics of terrestrial isopods show great diversity. Females in all species bear high maternal expenditures, whereas males invest mostly only in mating effort. Copulation costs were estimated and compared with females' costs of brood production. Since it requires many copulations in most species to fertilize a clutch, direct copulation costs are certainly not negligible in oniscidean males.

Interspecific differences are very pronounced with respect to direct energy investments in clutches. The lowest direct investment was found in *Hemilepistus* which, after delivery, carries out a very intensive and extended broodcare. We find the highest direct investments in those species which adopted another pathway to extended broodcare, by increasing their investments in a small number of progeny during pregnancy, then delivering large to extremely large young.

In aggregating species, favourable conditions for a resource defence polygyny of dominant males seem to exist but, on the grounds of sensory physiological constraints, these possibilities can only be very imperfectly exploited. Transient or permanent mate guarding is common in species digging burrows. This mate guarding is pre as well as postcopulatory.

In *Hemilepistus*, copulations in the middle of the extended mating phase are most important for securing reproductive success, as experiments with X-ray irradiated males and paternity determination by enzyme polymorphism have proven. There is no sperm precedence for either first or last matings. Sperm mixing takes place until a certain quantity of sperm has been stored. Females admit additional copulations, but this sperm is not used for fertilization. Multiple insemination could have been primarily developed as a means of reducing negative inbreeding effects. Secondarily, in the pair-forming species, it helps pair cohesion.

43

Males in *Hemilepistus* are totally unable to assess females before parturial moult with regard to their reproductive prospects. They do not discriminate against females with a full sperm store, and thus some males end up as pure stepfathers. Females of different species show either direct or indirect mate choice, preferring large males as mates; from these males both 'good genes' and direct benefits may be obtained.

To understand the many specialities with respect to sexual and reproductive tactics of the monogamous, semelparous *Hemilepistus*, with its extended biparental broodcare and highly evolved social behaviour, a comparison with the syntopic *Porcellio albinus* is very helpful. This comparison shows that the dissimilar substrates, on and in which these two species live, probably had a profound influence on shaping their greatly differing life history patterns.

ACKNOWLEDGEMENTS

This study was supported by the Deutsche Forschungsgemeinschaft (Li 150/ 6–10). I appreciate the help of many of my graduate students during field and laboratory work. Thanks are due to my technical assistant Elisabeth Barcsay and especially to my wife for her most valuable and continuous help during field work. I am grateful to Loretta Rott and to Anne Rasa for correcting and improving the English.

REFERENCES

Akahira, Y. (1956) The function of thoracic processes found in females of the common wood-louse, *Porcellio scaber. J. Fac. Sci. Hokkaido Univ., Ser. 6, Zoology,* **12**, 493–8.

Alexander, R.D. (1979) *Darwinism and Human Affairs.* Univ. of Washington Press, Seattle and London.

Alexander, R. and Noonan, K.M. (1979) Concealment of ovulation, parental care, and human social evolution. In: *Evolutionary Biology and Human Social Behaviour: An Anthropological Perspective,* (eds N.A. Chagnon and W.G. Irons), Duxbury Press, North Scituate, Mass., pp. 436–53.

Bateson, P.P.G. (1983) Optimal outbreeding. In: *Mate Choice* (ed. P.P.G. Bateson) Cambridge University Press, Cambridge, pp. 257–77.

Blaich, R. (1978) *Analytische Elektrophoreseverfahren.* Thieme Verlag, Stuttgart.

Bradbury, J.W. and Andersson, M.B. (eds) (1987) *Sexual Selection: Testing the Alternatives.* Dahlem Workshop Report. John Wiley and Sons, Chichester.

Charnov, E.L. (1982) *The Theory of Sex Allocation.* Princeton University Press, Princeton.

Cotelli, F., Ferraguti, M., Lanzavecchia, G. and Donin, C.L.L. (1976) The spermatozoon of peracarida. I. The spermatozoon of terrestrial isopods. *J. Ultrastr. Res.,* **55**, 378–90.

Dewsbury, D.A. (1982) Ejaculate cost and male choice. *Am. Natur.,* **119**, 601–10.

Edney, E.B. (1954) Woodlice and the land habitat. *Biol. Rev.,* **29**, 185–219.

References

Edney, E.B. (1968) Transition from water to land in isopod crustaceans. *Am. Zool.*, **8**, 309–26.

Emlen, T. and Oring, L.W. (1977) Ecology, sexual selection, and the evolution of mating systems. *Science*, **197**, 215–23.

Fain-Maurel, M.A., (1970) Le spermatozoide des isopodes. In *Comparative Spermatology* (ed. B. Baccetti), Academic Press, New York, pp 221–35.

Ferrara, F., Maschwitz, U., Steghaus-Kovac, S. and Taiti, B. The Genus *Exalloniscus Stebbing, 1911 (Crustacea, Oniscidea) and its relationship with social insects. Bull. Instituto Entomol. Univ. Parma* (in press).

Golla, A. (1984) Untersuchungen zum Fortpflanzungsverhalten und zum Geschlechterverhältnis der Landassel *Porcellio scaber* (Crustacea Isopoda Oniscoidea). *Diplomarbeit an der Julius-Maximilians-Universität, Würzburg.*

Gwynne, D.T. (1984), Male mating effort, confidence of paternity, and insect sperm competition. In: *Sperm Competition and the Evolution of Animal Mating Systems* (ed. R.L. Smith), Academic Press, New York, London, pp. 117–49.

Halliday, T.R. (1978) Sexual selection and mate choice. In: *Behavioural Ecology: An Evolutionary Approach* (eds J.R. Krebs and N.B. Davies), Blackwell Scientific Publications, Oxford, pp. 180–213.

Hamilton, W.D. (1967) Extraordinary sex ratios. *Science*, **156**, 477–88.

Hamilton, W.D. (1979) Wingless and fighting males in figwasps and other insects. In *Reproductive Competition and Sexual Selection in Insects* (eds M.S. Blum and N.A. Blum), Academic Press, New York, pp. 167–220.

Hamilton, W.D. and Zuk, M. (1982) Heritable true fitness and bright birds: a role for parasites? *Sciences*, **218**, 38–7.

Harris, H. and Hopkinson, D.A. (1976) *Handbook of Enzyme Electrophoresis in Human Genetics.* North-Holland Publishing Company, Amsterdam.

Hirschfield, M.F. and Tinkle, D.W. (1975) Natural selection and the evolution of reproductive effort. *Proc. Natl. Acad. Sci.*, **72**, 2227–31.

Hoese, B. (1984) The marsupium in terrestrial isopods. *Symp. Zool. Soc. Lond.*, **53**, 65–76.

Howard, R.D. (1978) The evolution of mating strategies in bullfrogs, *Rana catesbeana. Evolution*, **32**, 850–71.

Janssen, H.-H. and Hoese, B. Morphological and physiological studies on the marsupium in terrestrial isopods. In: *Proceedings of the Second Symposium on Biology of Terrestrial Isopods.* Monitore Zoologico Italiano, Monograph series, Urbino 1986 (in press).

Johnson, C. (1982) Multiple insemination and sperm storage in an isopod, *Venezillo evergladensis* Schultz, 1963. *Crustaceana*, **42**, 225–32.

Körner, R. (1986) Untersuchungen zum Reproduktionsverhalten und zur Populationsgenetik von *Hemilepistus reaumuri* Audouin & Savigny unter Anwendung der Zymogramtechnik. *Diplomarbeit an der Julius-Maximilians-Universität Würzburg.*

Linsenmair, K.E. (1972) Die Bedeutung familienspezifischer 'Abzeichen' für den Familienzusammenhalt bei der sozialen Wüstenassel *Hemilepistus reaumuri* Audouin u. Savigny (Crustacea, Isopoda, Oniscoidea). *Z. Tierpsych.*, **31**, 131–62.

Linsenmair, K.E. (1979) Untersuchungen zur Soziobiologie der Wüstenassel *Hemilepistus reaumuri* und verwandter Isopodenarten (Isopoda, Oniscoidea): Paarbindung und Evolution der Monogamie. *Verh. Dtsch. Zool. Ges.*, **1979**, 60–72.

Linsenmair, K.E. (1984) Comparative studies on the social behaviour of the desert isopod *Hemilepistus reaumuri* and of a *Porcellio* species. *Symp. Zool. Soc., Lond.*, **54**, 423–53.

Linsenmair, K.E. (1985) Individual and family recognition in subsocial arthropods, in particular in the desert isopod *Hemilepistus reaumuri*: In B. Hölldobler and M. Lindauer (eds), *Experimental Behavioural Ecology Sociobiology*, Symposium in Memoriam K.v. Frisch. *Fortschritte der Zoologie*, **31**, 411–36.

Linsenmair, K.E. (1987) Kin recognition in subsocial arthropods, in particular in the desert isopod *Hemilepistus reaumuri*. In *Kin Recognition in Animals* (eds D.J.L. Fletcher and C.D. Michener), J. Wiley and Sons, Chichester, pp. 121–208.

Linsenmair, K.E. and Linsenmair, Ch. (1971) Paarbildung und Paarzusammenhalt bei der monogamen Wüstenassel *Hemilepistus reaumuri* (Crustacea, Isopoda, Oniscoidea). *Z. Tierpsych.*, **29**, 134–55.

Lueken, W. (1962) Zur Spermienspeicherung bei Armadillidien (Isopoda Terrestria). *Crustaceana*, **5**, 27–34.

Lueken, W. (1968) Mehrmaliges Kopulieren von *Armadillidium* – Weibchen (Isopoda) während einer Parturialhäutung. *Crustaceana*, **14**, 113–18.

Maynard Smith, J. (1987) Sexual Selection – a classification of models. In: *Sexual Selection: Testing the Alternatives* (eds J.W. Bradbury and M. Andersson) Dahlem Workshop Report, John Wiley and Sons, Chichester, pp. 9–20.

Mead, F. (1967) Observations sur l'accouplement et la chevauchée nuptiale chez l'Isopode *Tylos latreillei* Audouin. *C.R. Acad. Sc. Fr.*, **264**, 2154–7.

Mead, F. (1976) La place de l'accouplement dans le cycle de reproduction des Isopodes terrestres (Oniscoidea). *Crustaceana*, **31**, 27–41.

Nakatsuru, K. and Kramer, D.L. (1982) Is sperm cheap? Limited male fertility and female choice in the lemon tetra (Pisces, Characidae). *Science*, **216**, 753–5.

Parker, S.A. (1970) Sperm competition and its evolutionary consequences in the insects. *Biol. Rev.*, **45**, 525–68.

Patané, L. (1940) Sulla struttura e le funzioni del marsupio di *Porcellio laevis* Latreille. *Archivo zool. ital.*, **28**, 271–96.

Ridley, M. and Thompson, D.J. (1979) Size and mating in *Asellus aquaticus*. *Z. Tierpsychol.*, **51**, 380–97.

Sassaman, C. (1978) Mating systems in porcellionid isopods: Multiple paternity and sperm mixing in *Porcellio scaber* Latr. *Heredity*, **41**, 385–97.

Schmalfuss, H. Revision der Landisopoden-Gattung *Porcellio* Lat. 3. Teil: Beschreibung von *P. linsenmairi* nov. spec. und Nachbeschreibung weiterer vier Arten aus Nord-Afrika (Isopoda, Oniscidea). *Spixiana*, (München) (Submitted).

Shine, R. and Bull, J.J. (1979) The evolution of live-bearing in lizards and snakes. *Am. Natur.*, **113**, 905–23.

Strassmann, B.I. (1981) Sexual selection, paternal care, and concealed ovulation in humans. *Ethol. Sociobiol.*, **2**, 31–40.

Sutton, S.L. (1968) The population dynamics of *Trichoniscus pusillus* and *Philoscia muscorum* (Crustaea, Oniscoidea) in limestone grassland. *J. Anim. Ecol.*, **37**, 425–44.

Sutton, S.L., Hassall, M., Willows, R., Davis, R.C., Grundy, A. and Sunderland, K.D. (1984) Life histories of terrestrial isopods: A study of intra- and interspecific variation. *Symp. zool. Soc. Lond.*, **53**, 269–94.

Tinkle, D.W. and Hadley, N.F. (1975) Lizard reproductive effort: calorific estimates and comments on its evolution. *Ecology*, **56**, 427–34.

Trivers, R.L. (1972) Parental investment and sexual selection: In *Sexual Selection and the Descent of Man, 1871–1971*, (ed B.G. Campbell), Aldine Press, Chicago, pp. 137–79.

Vandel, A. (1937) Recherches sur la sexualité des Isopodes: II. Les conditions de la fécondation chez *Trichoniscus provisorius. Bull. Biol. Fr. Belg.*, **71**, 206–19.

Vandel, A. (1941), Recherches sur la génétique et la sexualité des Isopodes terrestres. 2. Sur la longévité des spermatozoides à l'intérieur de l'ovaire d'*Armadillidium vulgare. Bull. Biol. Fr. Belg.*, **75**, 364–7.

Walker, W.F. (1980) Sperm utilization strategies in nonsocial insects. *Am. Natur.*, **115**, 780–99

Warburg, M. (1968) Behavioural adaptations of terrestrial isopods. *Am. Zool.*, **8**, 545–59.

Warburg, M. (1987) Isopods and their terrestrial environment. *Advanc. Ecol. Res.*, **17**, 187–242.

Warburg, M., Linsenmair, K.E. and Bercovitz, K. (1984) The effect of climate on the distribution and abundance of isopods. *Symp. Zool. Soc. Lond.*, **54**, 339–67.

Wickler, W. and Seibt, U. (1981) Monogamy in crustacea and man. *Z. Tierpsychol.*, **57**, 215–34.

Williams, G.C. (1966) *Adaptation and Natural Selection.* Princeton University Press, Princeton, N.J.

Wittenberger, J.F. (1979) The evolution of mating systems in birds and mammals. In: *Handbook of Behavioral Neurobiology, Vol. 3: Social Behavior and Communication*, (eds P. Marler and J. Vandenbergh), Plenum Press, New York, pp. 271–349.

Würsching, U. (1987) Reproduktionsbiologische und soziobiologische Untersuchungen an *Porcellio linsenmairi* (Isopoda, Porcellionidae). *Diplomarbeit an der Julius-Maximilians-Universität Würzburg.*

CHAPTER THREE

Mate guarding in geese: awaiting female receptivity, protection of paternity or support of female feeding?

Jürg Lamprecht

3.1 INTRODUCTION

In most animal species males can increase their reproductive success by copulating with more than one female, whereas there is usually no fitness gain for females copulating with more than one male. As sex ratios tend to be close to 1:1, this difference in reproductive potential underlies the widespread male competition for females.

Copulation may not occur when a male encounters a female. She may be unreceptive, and/or defended by another male. If undefended females are encountered frequently, the male's best reproductive strategy will be to go on searching for a receptive one; if not, the male should be ready to fight for a non-receptive female and defend her against rivals until he can inseminate her (Wickler and Seibt, 1981). Such *precopulatory mate guarding* occurs mainly in species in which fertile copulations are confined to a very short period of the female's reproductive cycle (Ridley, 1983; Parker, 1974; Wickler and Seibt, 1981; Grafen and Ridley, 1983).

In certain circumstances a selection pressure acts on males to prevent further mating of their females with others. This happens if females allow several copulations before fertilization of a batch of eggs is complete if sperm of the last copulation fertilizes at least some of the eggs (Compton *et al.*, 1978; Diesel, 1985) and if unfertilized females are not easily available. This favours the evolution of *postcopulatory mate guarding*, lasting until the end of the fertile period of the female's reproductive cycle (Birkhead, 1982; McKinney *et al.*, 1983, 1984).

In many species males do not stay with the female for the whole reproductive cycle. Here mate guarding is often easy to recognize, as defence of the female and maintenance of proximity (often in the form of 'clinging' to the female) have a definite start and end. When mates live together in harem groups or monogamous pairs over whole reproductive cycles, males may still need to prevent extra-marital copulations of their females, but their mate guarding is less conspicuous. Yet quantitative measures of inter-mate distance may show especially close mate-proximity and/or peak frequencies of aggressive behaviours during the mating period, with the male mainly responsible for both. Only to the degree that he actively promotes proximity and defends the female against rivals can he be said to 'guard' his female.

Where the function of mate guarding was studied in some detail, it always turned out that typical precopulatory mate guarding served to monopolize a female until she became receptive (Ridley and Thompson, 1979; Birkhead and Clarkson, 1980; Ward, 1983), while the role of postcopulatory mate guarding was likely to be paternity protection (Birkhead, 1979, 1982; Björklund and Westman, 1983; Birkhead *et al.*, 1985; Davies, 1985; Moller, 1985). Little is known about other functions of mate guarding during the mating period (Ashcroft, 1976; section 3.4).

This chapter demonstrates that bar-headed ganders (*Anser indicus*) guard their females during the pre-laying period. During this time of the year most copulations occur. Male removal experiments were performed and time budget data collected in an attempt to find the adaptive value of mate guarding in geese.

3.2 BIRDS AND GENERAL METHODS

Indian or bar-headed geese (*Anser indicus*) are the most common Asian geese. They winter in the Indian lowlands and breed in the highlands of Central Asia, north of the Himalayas. Their breeding colonies, with minimum distances of 2–3 m between nests, are often located on small islands (Dementiev and Gladkov, 1967; Kydyraliew, 1967; Schäfer, 1938).

Data was collected in a free-flying flock of about 100 birds living on the small lake (7.8 ha) of the Max-Planck Institute in Southern Germany which is fenced in against disturbance. Inside the fence some 1000 m^2 of grazing area is available, but additional ad lib pellet food is provided daily in a trough near the lake or in a cage with a trap door where the birds are caught for sexing and ringing in autumn. Permanent food supply and an ice-free section of lake keep most birds from leaving in winter. Breeding boxes (Lamprecht, 1986b; Würdinger, 1973) supplied with straw are provided on floating platforms (2.5 × 2.5 m) at the beginning of April. Nests are checked daily or at two-day intervals for data on egg laying, start of incubation, clutch size and egg loss.

The main bulk of data was collected from March 1983 until August 1984.

Additional observations and the male removal experiments were carried out in March and April 1985 and 1986. Methods are described in detail with the results of the different data sets.

Non-parametric statistical methods were used, and *P*-values are two-tailed unless otherwise stated.

3.3 RESULTS

3.3.1 Mate proximity and season

From August 1983 until August 1984 distances between mates in 31 well-established pairs were recorded five times a month (at about equal intervals). Each pair was located in turn and then watched for exactly one minute. At the end of this minute the following information was recorded: distance between mates, behaviour of each mate and location of pair (on land, on breeding platform). No records were made with both mates in the water, as the birds' leg rings were not visible. Some pairs disappeared from the flock during the year of study, other pairs formed (or came back) and were then included in the sample.

For the presentation in Figure 3.1 distances were excluded when a pair member was on a breeding platform or interacting socially with the mate (courtship), with the offspring (brooding) or another goose (aggression). Thus the data points represent distances on land during feeding, drinking, resting or preening. Mean inter-mate distances for each pair were calculated for each

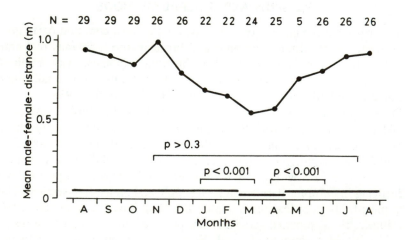

Figure 3.1 Mean inter-mate distances on land in the course of one year. N = number of pairs. Black horizontal bars = periods lumped for comparison with Sign tests.

month or longer intervals, and differences between periods tested with a Sign test based on the comparison of mean values of each pair.

Inter-mate distances shortened from August until March and then increased again (Figure 3.1). Differences between subsequent months tended to be insignificant, but mean distances from August–February and May–August were significantly greater (both $p < 0.001$, Sign test) than in March and April. There was no significant difference ($p > 0.2$) between the periods before and after March and April. The same significant picture emerged when March data alone were compared with distances in the other two periods.

First copulations were usually observed at the end of February or beginning of March; copulations were most frequent in March and early April, i.e. before the clutch was completed and incubation started (Dittami, 1981; and personal observation). Proximity between mates proved closest during the period of highest copulation frequency. Distances were measured for birds on land, while copulations occurred only in the water.

3.3.2 Contributions to proximity

To determine which mate is mainly responsible for maintaining proximity during the pre-laying and laying periods, the following data was collected during March and early April 1985.

Whenever a pair member moved away from the other, thereby definitely redirecting locomotion or switching to a new activity, it was noted whether this bird returned to the mate, whether it waited at some distance (often calling) until the other followed, or whether its mate followed promptly. Most of these events were observed in geese on land. From each record it was decided which bird had determined (initiated or 'allowed') the change in activity and location of the pair, and which had deferred to the other by following or returning. The records (2–9 per pair) were summarized for each of the 30 pairs.

In most pairs the male more often deferred to the female's activity than the female deferred to his ($p < 0.04$, Fisher exact probability test). On average the males deferred to their mate's activity in 66.3% of the records. Apparently males contribute more than females to inter-mate proximity during this time of the year.

This conforms with the behaviour of Bewick's swans (*Cygnus columbianus bewickii*), where males tend to lead the movements of the pair in autumn, but flight is more commonly initiated by females after 1 January (Rees, 1987).

3.3.3 Male removal experiments

If maintaining proximity to his female serves to protect the gander's paternity, then extra-pair copulations of females should be observed after removal of their males (Björklund and Westmann, 1983).

In early April 1985 we removed the males of 10 different pairs and observed the females during the next four hours. At the same time we counted every copulation occurring in the flock, which comprised 43 adult males and 57 adult females. As two females were monitored simultaneously, total flock observation time amounted to 20 h. The 62 copulations observed gave an average frequency of 0.054 copulations per adult female per hour.

None of the deprived females copulated. They uttered a moderate amount of distance calls, but were otherwise indistinguishable in behaviour and location from other females in the flock. One female was courted by a young gander for some time, but completely ignored him.

In March and April 1986 the males of eight pairs were removed for two hours and their females observed in the meantime. Not one was courted by or copulated with another male. Assuming the same average copulation frequency of females as in 1985 and combining the results of the two years, 3.024 copulations of the deprived females would have been expected in the 56 observation hours.

These calculations are probably unrealistic, as unpaired females may copulate less often than paired females. Yet the removal experiments show that a gander runs a very low risk of losing paternity or indeed his mate when he leaves the female even for several hours. This risk does not seem to justify the continuous maintenance of close proximity to his female.

3.3.4 Male and female time budgets in the mate guarding period

Male mate guarding, despite a very small risk of losing paternity, could have evolved if its costs are low. Some information on costs (or at least effort expenditure) can be gained by investigating male behaviour and time budget. Time budget data of 28 pairs in March and early April 1983 was obtained in the following way: three to five times a day, with intervals of at least one hour, each pair was located in turn and watched for exactly one minute, after which the activities of both mates were noted. During the following minute any new activities or events were noted for each adult. Sixty to 80 such observation minutes were collected for each pair. *Head up time* was expressed as the percentage of one-minute intervals starting with either 'head up' or 'extreme head up' (Lazarus and Inglis, 1978), *grazing time* as the percentage of intervals starting with grazing. The measures for *attacks* (fast approaches or bites) and for *threats* were the percentages of intervals containing at least one such event.

Figure 3.2 gives the means of all pairs. The males showed higher values for attacks, threats and head up time, but lower grazing time values than their females. All male–female differences were highly significant ($p < 0.001$, Sign test, $n = 28$ pairs).

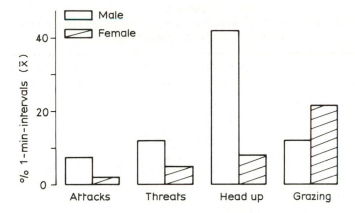

Figure 3.2 Mean percentage of one-minute periods containing at least one attack, at least one threat, or beginning with head up or with grazing. All male–female differences are significant ($p<0.001$ in Sign tests, $n = 28$ pairs).

The frequent attacks and threats, and the large amount of time spent with head up (negatively correlating with grazing time: section 3.3.5, Table 3.1), comprise a definite male effort and suggest substantial energy costs for males while mate guarding.

3.3.5 Correlation of time budget components with dominance and hatching success

If mate guarding had some other function than awaiting the female's receptivity, or protecting paternity after copulation, male behaviour during this period should still show a positive relationship with some aspect of male annual reproductive success.

To test this prediction the time budget components of pair members were correlated with each other; with subsequent hatching success (number of young leaving the nest in May 1983); and with pair dominance measured from 20 February to 10 April, 1983 (a pair's dominance value was the percentage of other pairs dominated of all those either dominant or subordinate: Lamprecht, 1986a). Spearman rank correlation coefficients are presented in Table 3.1. As hatching success and winter dominance values were available for only 25 pairs, correlations with these parameters are based on only $n = 25$. One-tailed significance tests for correlation coefficients were used.

The positive intercorrelations between male and female aggressive behaviour are not surprising, as mates are usually involved in the same agonistic episodes. More important for this study are the following results: none of the male and female time budget components correlated significantly with hatch-

Table 3.1

Spearman rank correlation coefficients between time budget components, winter dominance value and hatching success (HS)

		MA	MT	MH	MG	FA	FT	FH	FG	DOM	HS
Male											
Attacks	MA		0.71 ****	0.40 **	0.14	0.72 ****	0.36 *	0.28	0.29	0.63 ****	0.13
Threats	MT			0.44 **	−0.03	0.62 ****	0.20	0.19	0.37 *	0.60 ****	0.30
Head up	MH				−0.45 ***	0.23	0.17	0.15	0.25	0.32	0.25
Grazing	MG					0.23	0.12	0.20	0.30	0.10	0.09
Female											
Attacks	FA						0.36 *	0.45 ***	0.25	0.65 ****	0.16
Threats	FT							0.53 ****	0.14	0.56 ****	0.29
Head up	FH								− 0.08	0.56 ****	0.31
Grazing	FG									0.21	0.27
Dominance value	DOM										0.52 ****

* $p < 0.05$, ** $p < 0.025$, *** $p < 0.01$, **** $p < 0.005$ (one-tailed)
$n = 28$ (for correlations with DOM and HS, $n = 25$)

ing success of the pair. But male and female attacks and threats, and female head up time were positively related to winter dominance which indeed correlated positively with hatching success (Lamprecht, 1986b). Neither male nor female grazing time was related to dominance, but there was a significant positive correlation between male threats and female grazing time. Head up time in females was closely linked to aggressive behaviours, while male head up time showed a rather low correlation with aggressive behaviour but was negatively related to male grazing.

3.3.6 Recipients of the guarding male's aggression

If mate guarding served to defend the female against rivals, then attacks of the guarding male should be directed at males, but not at other females.

Males usually attack and threaten other males more often than other females (personal observations). But as paired males tend to position themselves between their mates and other geese, neighbours may more often attack them for their proximity rather than for their sex. This problem was circumvented by investigating (from 20 February to 10 April, 1983) whether unpaired females were more often displaced by the male or female of a pair. Instances were noted whenever the aggressive actions of pair members were clearly independent.

As paired males and females occur in equal numbers, single females should be displaced equally often by one or the other. This is to be expected if mates are equally aggressive and if there is no tendency to attack same-sexed conspecifics.

In 217 instances an unpaired female was displaced by a paired male, in 102 instances by a paired female. This is significantly different from a 1:1 expectation, i.e. 159.5:159.5 instances ($p < 0.001$, Chi-square test). As paired males generally seemed more aggressive than their females, other expected frequencies were calculated according to the mean differences (Figure 3.2) between mates in attack rate (expectation=254.5:64.5) and in threat rate (expectation=227.6:91.4). Chi-square tests showed that single females were displaced by paired males less often than expected from overall attack rate ($p < 0.001$), but not less often than expected from overall threat rate ($p > 0.15$).

Single females are at the bottom of the flock hierarchy (Lamprecht, 1986a), and no high-intensity attacks are needed to displace them. This may explain why they were less often displaced than the average goose is attacked. In terms of threats, however, paired males were apparently just as hostile to single females as they were to other birds in the flock, in spite of the fact that they sometimes copulate with single females (Lamprecht and Buhrow, 1987).

3.4 DISCUSSION

In section 3.1 two previously identified functions of guarding were mentioned: monopolizing the female until she becomes receptive (precopulatory mate guarding) and defending paternity (postcopulatory mate guarding). Both functions imply that in a gander's absence either his status as a paired male or his paternity is at stake, because the bereaved female may attach to or copulate with another male (Birkhead, 1981; Björklund and Westman, 1983). Yet the male removal experiments showed that a gander, even when absent for several hours, runs hardly any risk of losing his mate or paternity to another male. Mate guarding may have evolved in spite of a very low risk of cuckoldry if costs of guarding are also very low. But this is unlikely in view of the large amount of time spent on aggression and vigilance instead of feeding (Figure 3.1).

In a more mobile wild flock a male might lose contact with his mate when absent for too long. But this all-year problem of keeping the female in sight would not explain the extreme inter-mate proximity (Figure 3.1) and the peak frequencies of aggressive encounters (Dittami, 1981) in March and April only.

Howarth (1974) found in barnyard geese (*Anser anser*) that sperm in the female genital tract was able to fertilize eggs for less than ten days. And Elder and Weller (1954) had shown in domestic mallards (*Anas platyrhynchos*) that sperm viability dropped rapidly after five days. It seems safe to assume that in other geese also, sperm of one copulation cannot fertilize eggs laid more than two weeks later. In our bar-headed goose flock the first egg of the season was usually laid around mid-April (16.04.1983, 15.04.1984, 17.04.1985, 13.04.1986). Thus a paired male would lose nothing by allowing extra-pair copulations of his female until the end of March. Yet mate guarding is vigorous all through March. On their wintering grounds, wild barnacle geese (*Branta leucopsis*) spend most of the daytime grazing. Studies of these birds revealed that the frequency and intensity of aggressive acts (Black and Owen 1987) and the average distance between a pair and its nearest neighbours (Black, 1987) increased from January to April.

Mate guarding on land – where no copulations ever occurred – would be superfluous for preventing female extra-pair copulations. Forced copulations as reported for snow geese (*Anser c. caerulescens*) and Ross' geese (*Anser rossii*) (Mineau and Cooke, 1979) which happened on land (but usually on the nest) were never observed in bar-headed geese.

Both functions of mate guarding imply that the female is defended against rival males, so that the paired males' aggression against other (single) females, too, is not explained. Apparently, the characteristics of mate guarding in geese are not elucidated by factors related to sexual selection, i.e. to the probability of producing zygotes.

In various wild geese, female feeding time and nutrient reserves have proved important determinants of reproductive success. Heavier females produce larger clutches (Ankney and MacInnes, 1978) and/or incubate with fewer pauses and thus more successfully (Aldrich and Raveling, 1983; Prop *et al.*, 1984). The fact that paired female Canada geese (*Branta canadensis*) gain weight more rapidly than unpaired females before spring migration to the breeding grounds (McLandress and Raveling, 1981) suggests another function of mate guarding in geese, i.e. the protection of female feeding time and consequent build-up of nutrient reserves. Then mate guarding in geese would be unrelated to mate choice or male–male competition for females, and functionally similar to the *courtship feeding* in other birds, where the male, in presenting food to his female during the mating and laying period, enhances clutch size, hatching rate or both. Positive correlations of courtship feeding with the pair's reproductive success have been found in several species (Royama, 1966; Nisbet, 1977; Krebs, 1979).

As long as a male sires at least some of his female's offspring he benefits from actively increasing the female's reproductive success. A paired gander may therefore profit from boosting his female's food intake. Teunissen *et al.* (1985) studied the behaviour of individually ringed wild Brent geese (*Branta bernicla*) in spring prior to their departure for the breeding grounds, and counted the number of young they brought back in autumn. Pairs consisting of very aggressive and high-ranking males and females which spent much time feeding were the most successful.

Ashcroft (1976) showed that female eiders (*Somateria mollissima*) dabbling for mussels have a higher feeding rate when accompanied by a male warding off their conspecifics. McKinney *et al.* (1983) reported escorting and surveillance of the feeding female to be typical of male dabbling ducks (genus *Anas*) during the pre-laying and laying period. The authors conceded that mate-attendance and vigilance may protect the female from disturbance and allow more efficient feeding; but they also stressed that during this period when the female is fertilizable, males are also busy preventing forced copulations of their mates by other males.

Mate guarding may have various beneficial consequences. Yet when several selective forces act on a behaviour in the same direction, only one of them is usually responsible for the observed form or frequency (Curio, 1973). Neutralizing this one particular selective force is the behaviour's 'function in the strong sense' (Hinde, 1975) which should be distinguished from other beneficial consequences.

It is most likely that the ganders' mate guarding on land was shaped in evolution to serve the female's build-up of nutrient reserves. This function is compatible with all the results presented here, and it is supported by the correlation between male threat frequency and female grazing time (Table 3.1). This cannot be due to a general dominance effect because female grazing time was unrelated to dominance. The correlation suggests that mate guarding has a positive effect on female food intake. Yet grazing time did not correlate with hatching success. This is because in this flock, effects of nutrient reserves on reproductive success are masked by the readily available superabundant food supplies. Nevertheless, under natural conditions, mate guarding as practised by these semi-captive geese would enhance reproductive success.

3.5 SUMMARY

Mate guarding is shown on land by paired ganders in semi-captive bar-headed geese (*Anser indicus*) during the pre-laying and laying period (March and early April) when most copulations occur. Male removal experiments revealed a very low risk to the male of losing the female or copulations to rivals. Therefore the well-known functions of mate guarding – awaiting female receptivity during precopulatory mate guarding or protecting paternity in postcopulatory mate guarding – cannot account for the continuous close inter-mate proximity and vigorous mate defence on land where the guarding male wards off other males and females. It is more likely that mate guarding in geese serves to boost the female's build-up of nutrient reserves, which in field studies have proved an important determinant of a pair's reproductive success. A positive correlation between male threat frequency and female grazing time supports this hypothesis, which functionally relates mate guarding in geese to courtship feeding in other birds.

ACKNOWLEDGEMENTS

I am very grateful to H. Buhrow for her invaluable help in data collection, to T. Birkhead, J.M. Black and W. Wickler for constructive comments, to B. Knauer for drawing the graphs and to P. Rechten for correcting the English.

REFERENCES

Aldrich, T.W. and Raveling, D.G. (1983) Effects of experience and body weight on incubation behavior of Canada geese. *The Auk*, **100**, 670–9.

Ankney, C.D. and MacInnes, C.D. (1978) Nutrient reserves and reproductive performance of female Lesser snow geese. *The Auk*, **95**, 459–71.

Ashcroft, R.E. (1976) A function of the pairbond in the common eider. *Wildfowl*, **27**, 101–5.

Birkhead, T.R. (1979) Mate guarding in the magpie *Pica pica. Anim. Behav.*, **27**, 866–74.

Birkhead, T.R. (1981) Mate guarding in birds: conflicting interests of males and females. *Anim. Behav.*, **29**, 304–5.

Birkhead, T.R. (1982) Timing and duration of mate guarding in magpies, *Pica pica. Anim. Behav.*, **30**, 277–83.

Birkhead, T.R. and Clarkson, K. (1980) Mate selection and precopulatory guarding in *Gammarus pulex. Z. Tierpsychol.*, **52**, 365–80.

Birkhead, T.T., Johnson, S.D. and Nettleship, D.N. (1985) Extra-pair matings and mate guarding in the common murre *Uria aalge. Anim. Behav.*, **33**, 608–19.

Björklund, M. and Westman, B. (1983) Extra-pair copulations in the pied flycatcher (*Fidecula hypoleuca*). A removal experiment. *Behav. Ecol. Sociobiol.*, **13** 271–5.

Black, J.M. (1987) The pair-bond, agonistic behaviour and parent–offspring relationships in barnacle geese. PhD Thesis, University of Wales, Cardiff.

Black, J.M. and Owen, M. (1987) Variations in pair bond and agonistic behaviors in barnacle geese. In: *Wildfowl in Winter* (ed. M. Weller,). University of Minnesota Press, Minneapolis, pp. 23–38.

Compton, M.M., Van Krey, H.P. and Siegel, P.B. (1978) The filling and emptying of the uterovaginal sperm-host glands in the domestic hen. *Poult. Sci.*, **57**, 1696–700.

Curio, E. (1973) Towards a methodology of teleonomy. *Experientia*, **29**, 1045–58.

Davies, N.B. (1985) Co-operation and conflict among dunnocks, *Prunella modularis*, in a variable mating system. *Anim. Behav.*, **33**, 628–48.

References

Dementiev, G.P. and Gladkov, N.A. (1967) *Birds of the Soviet Union Vol. 4*, (Trans. from Russian by Israel Program Sci. Trans.), U.S. Dep. Inter. Natl. Sci. Found., Washington D.C.

Diesel, R. (1985) Fortpflanzungsstrategie und Spermienkonkurrenz der Seespinne *Inachus phalangium* (Decapoda, Maiidae). *Verh. Dtsch. Zool. Ges.*, **78**, 205.

Dittami, J.P. (1981) Seasonal changes in the behavior and plasma titers of various hormones in barheaded geese, *Anser indicus. Z. Tierpsychol.*, **55**, 289–324.

Elder, W.H. and Weller, M.W. (1954) Duration of fertility in the domestic mallard hen after isolation from the drake. *J. Wildl. Manage.*, **18**, 495–502.

Grafen, A. and Ridley, M. (1983) A model of mate guarding. *J. Theor. Biol.*, **102**, 549–67.

Hinde, R.A. (1975) The concept of function. In: *Function and Evolution in Behaviour* (eds G.P. Baerends, C. Beer, and A. Manning), Clarendon Press, Oxford, pp. 3–15.

Howarth, B.Jr. (1974) Sperm storage: as a function of the female reproductive tract. In: *The Oviduct and its Functions* (eds A.D. Johnson and C.W. Foley), Academic Press, New York, pp. 237–70.

Krebs, J.R. (1979) The efficiency of courtship feeding in the blue tit *Parus caeruleus. Ibis*, **112**, 108–10.

Kydyraliew, J. (1967) Die Streifengans (*Anser indicus*) im Tienschan Gebirge (trans. from Russian). *Ornithologija*, **8**, 245–53.

Lamprecht, J. (1986a) Structure and causation of the dominance hierarchy in a flock of bar-headed geese (*Anser indicus*). *Behaviour*, **96**, 28–48.

Lamprecht, J. (1986b) Social dominance and reproductive success in a goose flock (*Anser indicus*). *Behaviour*, **97**, 50–65.

Lamprecht, J. and Buhrow, H. (1987) Harem polygyny in bar-headed geese *Anser indicus. Ardea*, **75**, 285–92.

Lazarus, J. and Inglis, I.R. (1978) The breeding behaviour of the pink-footed goose: parental care and vigilant behaviour during the fledging period. *Behaviour*, **65**, 62–88.

McKinney, F., Derrickson, S.R. and Mineau, P. (1983) Forced copulation in waterfowl. *Behaviour*, **86**, 250–88.

McKinney, F., Cheng, K.M. and Bruggers, D.J. (1984) Sperm competition in apparently monogamous birds. In: *Sperm Competition and the Evolution of Animal Mating Systems* (ed. R.L. Smith), Academic Press, New York, pp. 523–45.

McLandress, M.R. and Raveling, D.G. (1981) Changes in diet and body composition of Canada geese before spring migration. *The Auk*, **98**, 65–79.

Mineau, P. and Cooke, F. (1979) Rape in the lesser snow goose. *Behaviour*, **70**, 280–91.

Moller, A.P. (1985) Mixed reproductive strategy and mate-guarding in a semi-colonial passerine, the swallow *Hirundo rustica. Behav. Ecol. Sociobiol.*, **17**, 401–8.

Nisbet, I.C.T. (1977) Courtship feeding and clutch size in common terns *Sterna hirundo*. In: *Evolutionary Ecology* (eds B Stonehouse, and C.M. Perrins), Macmillan, London, pp. 101–9.

Parker, G.A. (1974) Courtship persistence and female-guarding as male time investments strategies. *Behaviour*, **48**, 157–84.

Prop. J., van Eerden, M.R. and Drent, R.H. (1984) Reproduction success of the barnacle goose *Branta leucopsis* in relation to food exploitation on the breeding grounds, Western Spitsbergen. *Nor. Polarinst. Skr.*, **181**, 87–117.

Rees, E.C. (1987) Conflict of choice within pairs of Bewick's Swans regarding their migratory movement to and from the wintering grounds. *Anim. Behav.*, **35**, 1685–93.

Ridley, M. (1983) *The Explanation of Organic Diversity. The Comparative Method and Adaptations for Mating*, Oxford University Press, Oxford.

Ridley, M. and Thompson, D.J. (1979) Size and mating in *Asellus aquaticus*. *Z. Tierpsychol.*, **51**, 380–97.

Royama, T. (1966) A re-interpretation of courtship feeding. *Bird Study*, **13**, 116–29.

Schäfer, E. (1938) Ornithologische Ergebnisse zweier Forschungsreisen nach Tibet. *J. Orn.*, **86**, 1–49.

Teunissen, W., Spaans, B. and Drent, R. (1985) Breeding success in Brent in relation to individual feeding opportunities during spring staging in the Wadden Sea. *Ardea*, **73**, 109–19.

Ward, P.I. (1983) Advantages and a disadvantage of large size for male *Gammarus pulex* (Crustacea: Amphipoda). *Behav. Ecol. Sociobiol.*, **14**, 69–73.

Wickler, W. and Seibt, U. (1981) Monogamy in crustacea and man. *Z. Tierpsychol.*, **57**, 215–34.

Würdinger, I. (1973) Breeding of bar-headed geese (*Anser indicus*) in captivity. *Int. Zoo Yearb.*, **13**, 43–7.

CHAPTER FOUR

Helping in dwarf mongoose societies: an alternative reproductive strategy

Anne E. Rasa

4.1 INTRODUCTION

The majority of mammal species ensure the transmission of their genes to the subsequent generation by means of personal sexual reproduction. Each individual produced as a result of mating by a male and female of the same species carries one half of each of the parents' gene complement. These young thus have a coefficient of relatedness (r) of 0.5 with each of their parents and full siblings also bear a coefficient of relatedness of 0.5 to each other. This sociobiological premise has been used as an explanation for the presence of altruistic or helping behaviour in a wide variety of mammal, bird and insect species.

In closed societies, such as the social insects and some mammals, e.g. the naked mole rat *Heterocephalus glaber* (Jarvis, 1981), where each group contains only a single reproductive pair, the remaining group members being the offspring of these, helping can be considered as a means by which non-reproductive group members attain fitness by participating in raising their parents' young, ensuring that their genes are passed on to the subsequent generation in this way rather than by direct reproduction. This presupposes, however, that the presence of such helpers are essential to the young's survival. The majority of mammal species do not live in completely closed societies, although many of those in which helping has been reported consist of matrilineal groups e.g. lions (Bertram, 1975), warthogs (Mason, 1982), wild pig (Grundlach, 1968), elephants (Douglas-Hamilton, 1975). In some of these cases, males only become temporarily attached to the group during the breeding season. Curiously enough, helping appears to be less well developed or does not occur in other female-bonded social structures, such as the harem, e.g. several primate species such as langurs (Vogel, 1976), hamadryas baboons (Kummer, 1968), equines (Berger, 1977; Penzhorn, 1984).

A second major social structure in which helping has been recorded is the family group or extended family group. Family groups consist of a breeding pair and their offspring (of both sexes) over more than one breeding season. Extended family groups include F_2 offspring as well. The majority of family groups in which helping has been recorded are avian (Emlen and Vehren-kamp, 1983) and young may or may not disperse in the subsequent breeding season. In species where the young disperse, it is usually the sons which return the following breeding season to help raise their parent's new brood of chicks e.g. pied kingfisher (Reyer, 1985), white-fronted bee-eater (Emlen, 1981).

This phenomenon of sibling dispersal and return is far less frequently encountered in mammals, although it has been reported for the black-backed jackal, *Canis mesomelas* (Moehlman, 1979). The most common mode of help-ing in mammals is group or clan helping. These groups may be closed family systems, as in wolves (Mech, 1970) and banded mongooses (Sadie, personal communication) or open clans such as in brown hyenas (Mills, 1982). Some family-based groups in which a high degree of helping is observed, such as the Cape hunting dog (*Lycaon pictus*, Frame *et al.*, 1979) and the dwarf mongoose (*Helogale undulata*) are semi-open in that immigrants are accepted but only after long periods of 'trailing' (Rood, 1984; Rasa, 1987a). Here, however, acceptance into the group is not equated with reproductive success since only the dominant pair is able to reproduce successfully (Rasa, 1987a). In species such as the banded mongoose and brown hyena, several females in the group may reproduce during the same season and all group members act as helpers, whether they are related or not.

Helping, in the form of caring for the young, is a widespread phenomenon, predominantly amongst the carnivores within the Mammalia. This may be associated with the difficulty such species experience in obtaining food, not only for lactating females but also for the young. In addition, most carnivores must obtain their food at some distance from the den with the young, which exposes the latter to predation. This could be offset in two ways. Either one member of the group elects to forego foraging and to remain with the young while the mother searches for food to support her lactation ('babysitting') or other group members bring food to the mother and young at the den. In the dwarf mongoose, the species under consideration here, two different types of helping behaviour will be examined, together with their influence on infant survival rates, in an attempt to explain the evolution of helping in this species and its relationship to inclusive fitness.

4.2 METHODS

Field observations were conducted from January to April during 1980–1986 in the Taru desert, Kenya, this period coinciding with the birth of the young. Four focal groups with adjacent territories were observed from dawn to dusk daily for one month each. Groups are nomadic within territories containing 200+ termite mounds, the ventilation shafts of which are used as sleeping sites and shelters from predator attacks. Territory size varies from 0.65–0.96 km². Observations were made from a stationary vehicle using 12×50 binoculars and individuals were identified by means of natural cues, primarily facial characteristics and differences in pelage colour. Data were recorded on pre-prepared check sheets. Statistical tests used will be mentioned at appropriate points in the text.

4.3 RESULTS

Dwarf mongooses live in matriarchal family groups ranging in size from two to 32 individuals with a mean of 12.3. The group consists of the monogamous reproductive pair (alpha pair) and their offspring over several years, and occasional immigrants. Young animals are higher ranking than older ones and, in each age class, females are higher ranking than males (Rasa, 1972). Dwarf mongooses feed primarily on small arthropods which are a scattered food source requiring extensive search (Rasa, 1987a). Owing to this, there is no bringing of food to the young or the mother by other group members, as is typical for the majority of carnivores. The reasons why food provisioning does not occur in the Herpestinae have been elucidated by Ewer (1973). Helping in this species takes the form of protection of the young and food provisioning only during a later stage in their development. Protection takes two forms which will be dealt with separately.

4.3.1 BABYSITTING

Group members other than the mother have access to the litter immediately after birth. Babysitting consists of remaining with the young while the remainder of the group forages, and grooming and warming them while they are within the shelter of the termite mound. Babysitters remain with the young from the time the group leaves to forage until they return, usually at mid-day. During this period the babysitter is unable to feed. After the mid-day rest period, another individual takes over from the animal which has performed babysitting in the morning and the latter then joins the foragers in the afternoon. Occasionally the group does not return at mid-day. In this case, the babysitter remains with the young for the entire foraging period. Animals have been observed to engage in ritualized contests to determine which individual babysits first (Rasa, 1977) and there is a form of rota exchange between babysitters.

The young emerge from the mound and start accompanying the group at three and a half weeks of age. Since, prior to this emergence, the mongoose group retains its nomadic habits during moves (Rasa, 1987a), the young are transported between termite mounds by adult group members, predominantly the mother and babysitting siblings. During carrying, the young are grasped in the neck region on which they curl into a ball, the 'Tragestarre'. Once the young start accompanying the group, they are no longer carried continually and one youngster associates with one adult while it forages, and begs food from it. During the first ten days, the babysitters still carry the young across open areas or into hiding places should danger threaten. At the age of approximately five weeks, however, carrying ceases and the young are capable of keeping up with the adults at speed.

Weaning takes place between 40 and 45 days of age. The earliest food-begging observed was at 23 days of age, approximately the time when the young first start accompanying the babysitters. The young are capable of capturing and killing insect prey of their own between 28–30 days old but this prey capture is inefficient. Between the ages of three and a half and five and a half weeks of age, almost the entire solid food requirements of the young are met by the babysitters. Begging of food from adults continues up to approximately three and a half months of age but with a steadily decreasing success rate until food-begging finally ceases altogether and the young are capturing all their food themselves.

Babysitting therefore comprises all maternal functions with the exception of lactation. The mother suckles the young only at night or during the mid-day hours should the group return to the termite mound with the young. Babysitters are responsible for the entire care and feeding of the young, with the above exception, up to the age of approximately three and a half months. This time-period is comparable to that invested by the mother herself in her offspring (56 days' pregnancy plus 42 days' lactation = 3.2 months).

Not all members of the group babysit with equal frequency when babysitting is taken as the time spent with the young in the mound while the remainder of the group forages. Subordinate females, when they are present, appear to perform the majority of babysitting (Table 4.1). Should no subordinate females be present in the group, then their role is taken over by the males, including the father. The proportion of time the mother spends with the young instead of foraging also increases significantly in these cases. Although the sample size is too small to make any definite statement for groups with no subordinate females, the time invested by the mother appears to be independent of the number of males present.

Amongst the age and sex group performing the majority of babysitting, depending on group composition, there are also differences in the time

Table 4.1

The relationship between group size, number of subordinate females present and
percentage of time spent babysitting by males and females

Total no. adults in group	No. of subordinate females	Babysitting performed by subordinate females (%)	Babysitting performed by other group members (excluding mother) (%)
13	3	100	0
12	3	100	0
9	2	94.1	0
9	2	97.4	0
7	2	96.8	0
4	0	0	69.3
3	0	0	62.7
2	0	0	64.5

individuals spend babysitting. In groups with more than one subordinate
female, a single individual usually performs a disproportionately greater role
than other group members. In the groups with two subordinate females, the
highest ranking of these performed an average of 60.45% of the total babysitting, whereas in groups with three subordinate females, the highest ranking
ones performed an average of 43.6% of the total babysitting. Where only males
were present in subordinate roles, in one case two natal males performed
64.0% of the total babysitting, a single immigrant male 5.3%. In the case of
two natal males present, these shared 62.7% of the babysitting between them.

As previously mentioned, babysitting appears to be a high status behaviour
pattern since animals contest to perform it. This is reflected in the fact that,
when subordinate females are present, it is the highest ranking of these which
performs the majority of the behaviour pattern complex. These animals are
subadults of the age-class one to three years old.

4.3.2 GUARDING

Dwarf mongoose groups are exposed to a high degree of predator pressure,
especially from raptors (Rasa, 1983). An efficient form of coordinated vigilance behaviour has evolved to counteract this pressure (Rasa, 1986a). It is
the subordinate males in the group which are predominantly involved in this
activity, performing over 86.6% of total guarding. Guarding attains a high
degree of efficiency especially in larger groups, with over 90% of incipient
attacks by raptors being aborted by the guard seeing the predator first, warning, and the group going into hiding (Rasa, 1986a, 1987c).

Guarding behaviour is of particular importance during the period between emergence of the young from the mound to accompany the adult group members up until their independence of the babysitters. In larger groups, subordinate males may also act as guards at the mound with the babysitting subordinate female. Once the young leave the mound, however, not only are the mongooses vulnerable to a larger number of predator species, but they also are prey for smaller predators incapable of capturing an adult. The situation is exacerbated in groups with few or no subordinate females–males which would normally be involved in vigilance are acting as babysitters and accompanying the young while foraging. This reduces the number of animals available for guarding, and vigilance efficiency decreases. The effect of this decrease in vigilance efficiency is an increase in predation rate, especially of the young (Rasa, 1987a). Groups consisting of less than six adult group members over one year old lose all their young within the first year, 72.2% of these being killed within the first four months of life. In larger groups, however, where a sufficient number of animals are available to maintain a continuous and efficient vigilance system as well as sufficient individuals to perform babysitting, no young are lost to predators during a comparable stage in their development. Apart from having no increase in fitness, small groups also face extinction through predation (Rasa, 1986b). Predator pressure can thus be considered as the major factor predisposing toward group life in this species.

4.3.3 MALE AND FEMALE INVESTMENT

Both older male and female group members therefore play essential roles in the raising of their younger siblings, the former by affording them protection against predation, the latter by providing them with maternal care and nourishment with the exception of lactation. The comparative investment of each sex in the young is difficult to measure accurately. One indicator of investment, however, is the proportion of foraging time lost or reduction in food intake for both categories.

In groups of five or more animals, males performing vigilance behaviour forego an average of 20.5% of their total possible daily foraging time, since there is no significant difference in the frequency and duration of guarding between subordinate males (Rasa, in press). For babysitters, however, since it is one individual which predominantly stays with the young in the den prior to their emergence to accompany the group, the investment of subordinate females differs. If the female most active in this respect (the 'chief babysitter') is taken into account then, on an average, this individual loses approximately 52% of her total possible foraging time over a period of three and a half weeks. Since she is subsequently in constant attendance on one of the young from three and a half to five and a half weeks of age, the young obtains almost her entire food catch by begging it from her. This phase is followed by one of less intensive care for the subsequent eight weeks, after which the young are independent.

The actual investment in amount of food provided for the young during their period of accompanying a babysitter cannot be measured quantitatively under field conditions since food items vary in size and quality, and it is almost impossible to determine, especially at later stages of development, whether a youngster has received the food item, has been refused it, or has only received part of it. If investment is considered as the proportion of possible food intake lost, this would be equivalent to the foraging time lost during the period the young are immobile in the termite mound. For the purposes of this analysis, it will be assumed that the young receive 80% of the babysitter's catch between the ages of three and a half to five and a half weeks of age and that, subsequent to this, the amount of food received decreases linearly to 14 weeks of age. Based on these assumptions and including the foraging time lost when the young are immobile, investment for the 'chief' babysitter can be illustrated as shown in Figure 1. From this curve, measuring at half weekly intervals, she foregoes an average of 50.2% of her total possible food intake over the entire three and a half month period.

Lower ranking subordinate females, who do not take such an active part in babysitting during the period the young are in the mound have, however, the same demands placed on them by the young once these start accompanying the group. Based on their average contribution during the young's period of

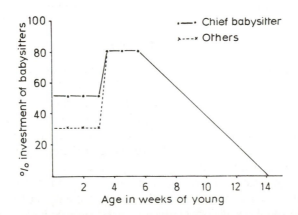

Figure 4.1 The estimated investment in food loss to babysitters during the period of the young's dependency, for the 'chief' babysitter and other subordinate females, where the litter size is a minimum of two individuals. Food loss in the first three and a half weeks is equated with time spent in the termite mound with the young, when the babysitter cannot forage. Subsequent to this, it is considered as the percentage of food captured which is presented to the young.

immobility and subsequent feeding of the young until three and a half months of age, their investment is almost equivalent to that of the 'chief' babysitter, at an estimated 45.3% of total possible food intake lost.

The estimation of the amount of food provided by the babysitters is based on weight gain recorded for hand-raised animals ($n = 9$) weaned at approximately the same age as observed in the wild and provisioned with solid food from 23 days of age onward. Based on the average weight gain in g/day it can be shown that 44.3 ±6.03% of the young's final weight at 38 days is gained during the first 24 days of life (milk diet only) while 57.48 ± 6.72% is gained during the subsequent 14 days (milk diet decreasing, provisioning with insects and meat). These data indicate that provisioning by the babysitters has a significant effect on the young's growth rate as measured by weight gain (Mann–Whitney U-Test, $p < 0.0001$). Since the mother is not involved in providing her young with insects, the entire body weight increase observed is associated with the solid food intake through provisioning by other group members.

Babysitting therefore exacts a higher investment over a shorter timespan than does vigilance behaviour. If a single litter is produced per year, then the average investment of the 'chief' babysitter over the entire year can be calculated as 14.6% of total intake lost, that for other babysitters as 13.2%. Should two litters per year be produced, as is the case when adequate rainfall occurs during the spring months, then the total annual investment is doubled, with 29.2% for the 'chief' babysitter and 26.4% for subordinate ones. These figures compare well with the average investment of males over the same timespan, which is 20.5% of their total possible foraging time spent guarding.

4.4 DISCUSSION

Reproductive altruism (Trivers, 1985) is defined as the situation in which an individual foregoes personal reproduction – in some cases for its entire life-span – in order to aid reproduction in others, especially if the individuals benefitted are relatives. The advantages of such reproductive altruism to the helping individual would be long-term investments such as inheritance of a territory in which it could subsequently breed itself or individuals which would act as helpers in their turn if the original helper breeds.

In the dwarf mongoose, however, group life in a group consisting of more then five adults has been shown to be obligatory, not only for fitness increase but also for personal survival since predator pressure is so high. In addition to territorial inheritance and reciprocal helping, personal survival must also be considered as a major and immediate factor predisposing towards the evolution of altruism, at least in this species. It has been shown that one of the helping roles, vigilance behaviour, is the single most important factor reducing mortality in the group, especially juvenile mortality (Rasa, 1987b, 1987c).

Discussion

The second altruistic role, babysitting, is exclusively young-oriented and seasonal while vigilance behaviour is continuous and group-oriented rather than directed towards one age or sex class. Both roles, however, are essential for efficient reproduction and, since the group consists of close relatives, the inclusive fitness of all group members.

Although the two roles are performed across different timespans, the estimated investment for both of them is approximately equal when averaged across the entire year. Investment was considered as the estimated food intake lost to the individual while performing the role. Since the subordinates have no reproductive success within the group (Rasa, 1987a), practically no chance of survival if they leave the group to form a splinter group of their own (Rasa 1986b), and low chance of reproductive success if they manage to attach themselves to another group (Rasa 1987a), then remaining within the natal group is a 'safe' way of increasing fitness with minimal risk and maximum benefits. This is especially true if the helper is a full sibling to the young being helped, which is true in the majority of cases. From a genetic point of view, this individual would increase its own fitness at the same rate of gene transference as if it were producing young of its own. For the majority of dwarf mongoose individuals, this is the only option left to gain fitness during their lifetimes.

Emlen (1982) has postulated for birds that ecological constraints can severely limit the possibility of personal reproduction and predispose toward the evolution of helping or cooperative breeding. The first of the two types of constraints he hypothesizes, the practical non-existence of breeding openings, is known to be a factor limiting reproduction in dwarf mongooses. This is the major constraint on species inhabiting stable, predictable environments, such as that of *Helogale*. His suggestion, that the best strategy for a non-breeder is to wait in an area of proven quality in the company of close kin until the individual attains sufficient age, experience and status to breed independently, appears to apply for this species. This is especially important since each individual gains an increment in inclusive fitness with each relative it helps raise. As the data have shown, without such helpers, the fitness of a group is zero and a minimum of four helpers are necessary for a gain in fitness. In contrast to bird species, where food limitation appears to be the major factor contributing to increased fitness, in dwarf mongooses, protection of the young also plays a major role.

The evolution of helping in dwarf mongooses therefore appears to be associated with two factors, predator pressure and mode of nutrition. The presence of a high predator pressure has necessitated the evolution of a helping role to compensate for this. Similar behaviour patterns have evolved in other small, diurnal, group-living mammals e.g. Belding's ground squirrel, *Spermophilus beldingi* (Sherman, 1977), marmots (Barash, 1974) and banded mongooses (personal observation). Whether, however, this behaviour is a

69

prerequisite of group life or has evolved as a compensatory factor for group living, probably differs with the species under consideration. For dwarf mongooses, however, it not only enhances the survival of adult group members but is essential for the inclusive fitness of the group as a whole.

The mode of nutrition of dwarf mongooses has also had far-reaching effects on behaviour and the evolution of sociality in this species, mainly because of the size of prey taken. Dwarf mongooses show differences in maternal behaviour and juvenile development compared with other carnivores, these probably being adaptations to the species' mode of nutrition and life style. Owing to the fact that they are primarily insectivorous, the typical helping behaviour observed in other carnivores, that of provisioning the mother and/or young in the den and an extended den-life for the young, are absent. Emphasis has been placed in ontogeny on the rapid maturation of behaviour patterns concerned with mobility and spatial orientation. As soon as these behaviour patterns have matured, at the age of three and a half weeks, and the young are able, they start accompanying the adults even though they are physically and behaviourally immature in other respects (Rasa, 1985). This is the shortest known den-life for any carnivore species to date. The necessity for such a strategy is obvious for, should the young restrict the group's movements over a prolonged period of time, then the resources available in the immediate area would be severely depleted since this must support not only the mother and young, but the entire group as well.

In addition to accelerated ontogeny, this mode of nutrition has also imposed constraints on the mother since the food items are too small to be transported to her at the den. She must search for them herself, thus leaving the young exposed to predation if no other strategy had evolved to compensate for this. Protection of the abandoned young is all the more important in this species since there is no permanent den or nest where the young can be hidden, since these are regularly moved with the group. It is the subordinate females of the group, predominantly, which forego their own foraging to remain with the young, thus ensuring that the mother receives sufficient nourishment to support lactation. The lactation period of 40–45 days is equivalent to that of other small carnivores of approximately the same body size e.g. *Mustela erminea* (Erlinge, 1977), *Suricata suricatta* (Ewer, 1963). For almost half this time, however, the young are receiving solid food from other group members, so the demands on the mother start being reduced earlier than for solitary species such as *Mustela* where the mother must also provide the solid nourishment for the young as well.

Helpers thus aid not only by increasing the survival rate of the young but also by decreasing the demands made by the young on the mother. This is especially important for a species feeding on very small prey items since it would be practically impossible for the mother alone to provision all litter members adequately, a problem which does not occur in carnivores which are large prey killers. The presence of several provisioning helpers serves to circumvent this bottleneck.

Dwarf mongooses appear to be unique amongst the carnivore species where helping has been recorded in that there is a division of labour in helping between the sexes. It is the male siblings which are primarily involved in protective functions, the females in food provisioning. The relative investment of both sexes, however, appears to be approximately equal reckoned on an annual basis, although female investment is very high when dependent young are present and active babysitters can be seen to lose condition. Once the young become independent, however, their investment is minimal and condition is quickly regained. For guards, however, where investment is low but steady across the whole year, such loss of condition has not been observed.

It has been shown that groups consisting of five or fewer adults have no reproductive success. This indicates that a minimum of four helpers are necessary to ensure the fitness of the group as a whole. In successful groups, if the average number of young surviving to four months of age is taken as an indicator of fitness (assuming that until this time the young are still receiving a direct investment from their siblings) then the average annual fitness for a group would be 2.44 ± 0.93 young/year ($n = 11$). If two litters are born per year, then this figure would be doubled. For a babysitter which starts babysitting at one year old (the same age at which the guards become active participants in vigilance behaviour), fitness gain during its lifetime may be appreciable, even though it does not breed itself, since dwarf mongooses have lifespans of up to 18 years. It would thus be an optimal strategy for a sibling to remain in its natal group for maximal fitness gain. This strategy, however, does not preclude personal reproduction if the reproductive animal of the same sex is lost or killed (Rasa, 1987a). This may explain why individuals remain with their natal groups for their entire lifespans (one known individual remained as a helper for 16 years), since the pay-offs in fitness gain by being a helper greatly exceed those of alternative strategies, except in instances where an individual leaves its own group to become a breeder in another group immediately. The chances of this occurring, however, are extremely small and it has never been observed in either wild or captive groups during 16 years observation.

Dwarf mongooses therefore appear to have adopted two different and interdependent strategies for gene transmission. The more common method of personal reproduction is reserved for only two members of the group, the dominant animals. All other group members, depending on their degree of relatedness to these two – which is, in general, high – increase their fitness through two sex-biased helping roles which require approximately the same amount of investment. These roles have been shown to be crucial for the survival of the young, hence fitness in this species is a direct correlate of reproductive altruism.

71

ACKNOWLEDGEMENTS

This study was supported by a grant from the Deutsche Forchungsgemeinschaft to the author. I am extremely grateful to Mr Ray Mayers for allowing me to conduct my studies on his ranch in Kenya.

REFERENCES

Barash, D.P. (1974) The evolution of marmot societies: a general theory. *Science,* **185,** 415–20.

Berger, J. (1977) Organisational systems and dominance in feral horses in the Grand Canyon. *Behav. Ecol. Sociobiol.,* **2,** 131–46.

Bertram, B.C.R. (1975) Social factors influencing reproduction in wild lions. *J. Zool.,* **177,** 463–82.

Douglas-Hamilton, I. (1975) *Among the Elephants.* Collins and Harvill Press, London.

Emlen, S.T. (1981) Altruism, kinship and reciprocity in the white-fronted bee-eater. In: *Natural Selection and Social Behaviour: Recent Research and New Theory* (eds R.D. Alexander and D. Tinkle). Chiron Press, New York, pp. 217–230.

Emlen, S.T. (1982) The evolution of helping. I. An ecological constraints model. *Am. Natur.,* **119,** 29–39.

Emlen, S.T. and Vehrenkamp, S.L. (1983) Co-operative breeding strategies among birds. In *Perspectives in Ornithology,* (eds A.H. Brush and G.A. Clark, Jr.), Cambridge University Press, Cambridge, pp. 93–120.

Erlinge, S. (1977) Spacing strategy in stoat, *Mustela erminea. Oikos,* **28,** 81–98.

Ewer, R.F. (1963) The behaviour of the meerkat, *Suricata suricatta. Z. Tierpsychol.,* **20,** 570–607.

Ewer, R.F. (1973) *The Carnivores.* Cornell University Press, New York.

Frame, L.H., Malcom, J.R., Frame, G.W. and van Lawick, H. (1979) Social organization in African wild dogs (*Lycaon pictus*) on the Serengeti plains, Tanzania, 1967–1978. *Z. Tierpsychol.,* **50,** 225–49.

Grundlach, H. (1968) Brutforsorge, Brutpflege, Verhaltens-ontogenese und Tagesperiodik beim Europaischen Wildschwein (*Sus scrofa* L.). *Z. Tierpsychol.,* **25,** 955–95.

Jarvis, J.U.M. (1981) Eusociality in a mammal: cooperative breeding in naked mole-rat colonies. *Science,* **212,** 571–3.

Kummer, H. (1968) *Social organization in Hamadryas baboons – a field study.* Karger Verlag, Basel.

Mason, D. (1982) Studies on the biology and ecology of the warthog (*Phacochoerus aethiopicus sundevalli* Lönnberg 1908) in Zululand. *D.Sc. Thesis,* University of Pretoria, R.S.A.

Mech, L.D. (1970) *The Wolf.* Natural History Press, New York.

Mills, M.G.L. (1982) The mating system of the brown hyaena, *Hyaena brunnea,* in the Southern Kalahari. *Behav. Ecol. Sociobiol.,* **10,** 131–6.

Moehlman, P.D. (1979) Jackal helpers and pup survival. *Nature,* **277,** 382–3.

Penzhorn, B. (1984) A long-term study of social organization and behaviour of Cape mountain zebras. *Z. Tierpsychol.* **64,** 97–146.

References

Rasa, A.E. (1972) Aspects of social organization in captive dwarf mongooses. *J. Mammal.*, **53**, 181–5.

Rasa, A.E. (1977) The ethology and sociology of the dwarf mongoose (*Helogale undulata rufula*). *Z. Tierpsychol.*, **43**, 337–407.

Rasa, A.E. (1983) Dwarf mongoose and hornbill mutualism in the Taru Desert, Kenya. *Behav. Ecol. Sociobiol.*, **12**, pp. 181–90.

Rasa, A.E. (1985) Sozialisation bei Zwergmangusten. In: *Verhaltensbiologie* (ed. D. Franck), Stuttgart, Thieme Verlag, pp. 251–7.

Rasa, A.E. (1986a) Coordinated vigilance in dwarf mongoose family groups: the Watchman's song hypothesis and the costs of guarding. *Ethology*, **71**, 340–4.

Rasa, A.E. (1986b) Ecological factors and their relationship to group size, mortality and behaviour in the dwarf mongoose *Helogale undulata* (Peters, 1852). *Cimbebasia*, **8**, 15–21.

Rasa, A.E. (1987a) The dwarf mongoose: study of behaviour and social structure in relation to ecology in a small, social carnivore. *Advances in the Study of Animal Behaviour. Vol. 17*, (ed. J. Rosenblatt), Academic Press, Petulama, California, pp. 121–63.

Rasa, A.E. (1987b) Sociability for survival: why dwarf mongooses live in groups. In: *Readings from the 19th International Ethology Conference, Vol. 4, Behavioural Ecology and Population Biology*, (ed. L.C. Drickamer), Privat Press, Toulouse, France, pp. 35–9.

Rasa, A.E. (1987c) Vigilance behaviour in dwarf mongooses: is it selfish or altruistic? *S.A.J. Zool.*, **83**, 587–90.

Reyer, H-U. (1985) Brutpflegehelfer beim Graufischer. In: *Verhaltensbiologie* (ed. D. Franck), Thieme Verlag, Stuttgart, pp. 277–82.

Rood, J. (1984) Pack life of the dwarf mongoose. In: *The Encyclopaedia of Mammals 1*, (ed. D.W. McDonald), Allen and Unwin, London, pp. 152–3.

Sherman, P.W. (1977) Nepotism and the evolution of alarm calls. *Anim. Behav.*, **28**, 1070–94.

Trivers, R. (1985) *Social Evolution*. Benjamin/Cummins Publishing Company, Menlo Park, California.

Vogel, C. (1976) Ökologie, Lebensweise und Sozialverhalten der Grauen Languren in verschiedenen Biotopen Indiens. *Z. Tierpsychol., Beiheft 17*, Parey Verlag, Berlin.

CHAPTER FIVE

Reproductive strategies of female gelada baboons

Robin I.M. Dunbar

5.1 INTRODUCTION

Gelada baboons (*Theropithecus gelada*) live in small reproductive units that conventionally contain a single breeding male with up to 12 post-puberty females and their associated dependent young. A number of these reproductive units share a common ranging area and, although the relationships between units are quite elastic, the units that share a common area spend most of their time associating with each other rather than with units that normally range elsewhere. These super-groups are termed bands (Kawai *et al.*, 1983). Within the reproductive units, females form closely bonded relationships among themselves (Dunbar, 1979, 1984). A detailed analysis of the functional significance of these relationships (Dunbar, 1984) suggests that they are designed to buffer the females against harassment and stress generated by the close proximity of other females in the same group.

This explanation for the structure of relationships among gelada females involves a three-step argument: (1) that harassment by other females of the same group causes low-ranking females to become infertile, (2) that, as a result of this, there is a negative relationship between fecundity and dominance rank and, finally, (3) that females try to buffer themselves against this loss of fertility by forming alliances amongst themselves. It is important to note that each of these three steps is an independent observational statement, though they are interpreted as being causally related. The relationship between dominance rank and fecundity, for example, could be due to any one of several different causes, though the claim is that there is evidence to attribute it to the effects of stress in this case.

The original analyses on which this explanation was based remain subject to some ambiguities. In particular, the data on which (2) was based do not show a statistically significant relationship between dominance rank and reproductive output. Moreover, it can be argued that even if the relationship does exist, then the cause may lie not in the reproductive suppression of low-ranking females but in their exclusion from the best food sources. I suspect that one of the reasons why the argument seems to be weak is that the relevant evidence is scattered through several different publications. On the grounds that individual criticisms carry weight only when sets of data are viewed in isolation, I want to set out the argument and the evidence in a coherent integrated form in order to try to show that, when the battery of evidence is taken together, my interpretation is more robust than appears at first sight. To do this, I shall both explain the logic of the original analyses in some detail (and I hope more clearly), then undertake further analyses of the same data sets and finally present some entirely new data bearing on the same issues. In the process, I shall try to show that these new analyses resolve some ambiguities and uncertainties in the original analyses.

The analyses presented here are based on data obtained from the population of gelada baboons living in the Sankaber area of the Simen Mountains National Park, Ethiopia, during 1971–2 and 1974–5. Details of both the animals and the methods of data sampling are given in Dunbar (1980a, 1980b, 1983a), so I shall not repeat them here. I also make use of the data from studies of gelada in the Gich area of the Simen carried out by Kawai and his co-workers (Kawai, 1979).

5.2 DOMINANCE AND REPRODUCTIVE SUCCESS

In the original paper (Dunbar, 1980a), the number of offspring less than 4.25 years old that each of 55 reproductive females had associated with them was used to estimate the relationship between fecundity (birth rate) and dominance rank. The data derived from 12 reproductive units in one band studied at Sankaber during 1974–5. I tried to show that although the relationship was weak on face value, nonetheless the only plausible explanation for the data was that fecundity was negatively related to rank. I did this in two ways. First, I tested a set of alternative hypotheses that might have accounted for the observed distribution. These alternatives were: (1) that differences in age-specific fecundity would produce just such a negative relationship if dominance rank was correlated with age; (2) that a density-dependent effect acting equally on all females would produce a weak negative relationship since low-ranking females can only come from large units; (3) that family size determined dominance rank rather than the other way around; and (4) that dominant females attract immatures (especially orphans) who associate with them in order to benefit from their support during agonistic encounters. Each of these was excluded by a separate analysis of relevant data (Dunbar, 1980a).

I then went on to test the relationship itself by using it to predict a relationship between total number of offspring and the number of females in units of two bands censused in 1971. During the course of this analysis, I also tested two other null hypotheses: that the observed distribution could be explained on the assumption either (1) that all females reproduced at the same average rate or (2) that they all reproduced at the same rate as the maximum achieved by the females of the 1974 sample. This analysis demonstrated (a) that the slope parameter of the regression equation for numbers of offspring in the 1971 sample did not differ significantly from that predicted by the relationship between dominance rank and birth rate and (b) that the slope parameter was significantly lower than both of those predicted by the two null hypotheses that birth rate was constant and independent of both group size and dominance rank.

I accept that the evidence in favour of the hypothesis for an inverse relationship between reproductive output and rank might be weak, not least because the slope of the equation for reproductive output regressed on rank is not itself significantly different from zero owing to considerable variance in the reproduction outputs of individual females. However, we can strengthen our belief in the inferred relationship in several different ways. First, we can show that we consistently obtain a negative slope of about the same magnitude from different sets of data. Secondly, we can point to the fact that this initial analysis only provides an approximate estimate of birth rates because various other factors that also influence birth rates introduce a great deal of error into the data. One of these factors is the animals' own attempts to use social tactics (e.g. coalitions) to buffer themselves against the worst of the effects caused by low rank. We need to appreciate that what we observe in the field is not just the effects of a single variable like dominance rank, but the net result of this phenomenon combined with the animals' attempts to circumvent the reproductive disadvantages that it imposes on them. Our problem, in a sense, is to be able to pare away these overlying complications in order to reveal the underlying phenomenon in undistorted form. This is clearly no easy task, but we have to take the problem seriously for it would be naive to assume that animals as intelligent as primates might be prepared to accept such costs without making some attempt to minimize their impact. If we remove the effects of any of these confounding variables, then we should find that the slope of the relationship becomes steeper. Finally, we can try to adduce evidence for a real physiological mechanism that could produce the hypothesized relationship between birth rate and rank. If we can do this, then any attempt to discredit the relationship between birth rate and rank will face the secondary problem of having to explain why we should find evidence for the physiological mechanism but none for its effect on fecundity.

5.2.1 Alternative tests of the hypothesis

So far, five independent tests of the hypothesized relationship between rank and fecundity have been carried out. These are: (1) the original analysis of numbers of offspring plotted against dominance rank for 55 females of the Main band, Sankaber, in 1974–5; (2) an analysis of the total number of immatures in relation to the number of females in 30 units of the Main and Abyss bands at Sankaber in 1971 (a total of 117 reproductive females); (3) a similar analysis for the 29 units of the E, F and K bands at Gich in 1973 (a total of 125 females); (4) an analysis of the mean birth rate for a complete band plotted against the mean rank of all the females in it for five bands (a total of 288 females); and (5) an analysis of actual birth rates plotted against dominance rank for the 55 females of the Main band, Sankaber, during 1974–5.

Each of these five analyses yields a negative slope for fecundity plotted against rank (Table 5.1). The likelihood of obtaining five negative slopes by chance alone is $p=0.031$ (1-tailed binomial test). We can make a stronger test of the hypothesis by using Fisher's procedure for combining significance levels from independent tests of the same hypothesis (Sokal and Rolf, 1969) to ask how likely it is that we would obtain a distribution of p values as extreme as these by chance if there were no underlying relationship between fecundity and rank. This yields $\chi^2 = 18.783$ (d.f.$=2k=10$, $p<0.05$) which implies substance to the underlying trend. Note also that the slope parameters of four of the five equations are very similar to each other, despite the different data bases.

5.2.2 Removing confounding variables

The original analysis sought to estimate birth rates from data on net reproductive output over a 4.25-year period. There are at least four factors that are likely to affect birth rates which will introduce randomizing error into these data, so increasing the variance. These are: (1) fecundity varies with a female's age (Dunbar, 1980b); (2) not all the females in the sample would have been reproductively active for the full 4.25 years over which the sample was taken; (3) birth rates vary from year to year in response to extrinsic environmental variables (Dunbar, 1980b); and (4) a female's dominance rank within her unit is likely to vary considerably with time, so that her total reproductive output is a function not of her current rank but of her average rank over the five years since the oldest offspring was conceived.

Since rank is not a function of age among gelada females (Dunbar, 1980a), the pooling of data from both young and old females will obviously increase the variance in the data, even though all females are equally affected by rank-dependent effects. If the data for each age class are analysed separately, we

Table 5.1

Tests of the relationship between a female's fecundity and her dominance rank within her unit

Test	Slope[a]	t	d.f.	p (1-tailed)[b]	Source
1. Total offspring[c] vs dominance rank (Sankaber, 1974)	−0.028	−0.658	53	0.257	Dunbar, 1980a (fig. 6)
2. Total offspring[d] vs group size (Sankaber, 1971)	−0.026[f]	−0.448	28	0.329	Dunbar, 1980a (fig. 7)
3. Total offspring[d] vs group size (Gich, 1973)	−0.006[f]	−0.075	27	0.470	Dunbar, 1984 (fig. 69)
4. Mean birth rate vs mean rank[e] (5 bands, Sankaber)	−0.048	−1.919	3	0.075	Dunbar, 1984 (table 6)
5. Birth rate vs dominance rank (Sankaber, 1974)	−0.042	−11.141	1	0.028	Dunbar, 1986 (table 15.1)

[a] Slope parameter for linear regression fitted to data by least squares.
[b] Probability that slope departs from zero in a negative direction only.
[c] Total number of living offspring less than 4.25 years old belonging to each female.
[d] Total number of living offspring less than 4.25 years old in the whole unit (including orphans).
[e] Mean rank of females within their own units for all the females in a given band.
[f] To maintain comparability for present purposes, the mean offspring per female was regressed on group size using the original data (2 bands at Sankaber, 3 bands at Gich).

Table 5.2

Regression equations for the number of offspring less than 4.25 years old regressed on the female's dominance rank for females of 11 reproductive units censused at Sankaber, Simen, during 1974–75

Age class of female	Regression equation	r^2	n	$t (b = 0)$	p (1-tailed)
Subadults (4–5 years)	$y = 3.076 - 0.258x$	0.077	18	−1.152	0.133
Young adults (6–8 years)	$y = 1.976 - 0.111x$	0.074	12	−0.979	0.173
Old adults (9+ years)	$y = 2.143 - 0.101x$	0.078	22	−1.369	0.092

find that the slope parameter for number of offspring regressed on rank is negative in all three cases (Table 5.2). More importantly, all three slope parameters are steeper than when all age classes are pooled together. Treating each age class as an independent test of the hypothesis, Fisher's procedure indicates that, taken together, the slope parameters do differ significantly from zero ($\chi^2 = 12.214$, d.f.$= 2k = 6$, $p = 0.05$). Note that subadults seem to be more seriously affected by the rank-related factor than are adult females.

The second problem arises from the fact that although we are considering a sample period equivalent to the previous 4.25 years, the younger females in the sample will not have been reproductively active for the whole of this period. Indeed, with a relatively long mean interbirth interval of 2.14 years, many of the subadult females will barely have had time to produce even a single infant. Table 5.2 indicates that subadult females do indeed behave differently from adult females. In contrast, the two older age classes have regression slopes and origins that are virtually identical. Including data from all three classes is bound to increase the variance by a significant proportion.

The third source of randomizing error is the fact that birth rates vary considerably from year to year due to fluctuating weather conditions (Dunbar, 1980b). This effect is compounded by the fact that a female's dominance rank is unlikely to remain constant over so long a period of time. We can remove the influence of both of these effects by considering actual birth rates within a defined period of time (e.g. a single year). Table 5.1 demonstrates that doing so yields very consistent results: tests (4) and (5) are both individually significant and their slope parameters are virtually identical.

In addition to the randomizing effects of these factors, there is one other feature of gelada biology that is likely to increase the variance in the data, namely the fact that at least one female in each unit is able to use the male as a buffer against harassment by other females even though, by doing so, she cannot improve her own dominance rank within the unit (Dunbar 1980a, 1983b). Of 11 females who were the male's main social partner within their units, eight had more offspring than a non-partner female matched for age, rank and harem size: on average, partner females had 8.3% more offspring than matched non-partner females (Dunbar, 1984). Some low-ranking females will consequently have more offspring than they ought to if reproductive suppression was acting in an unfettered way. It is also worth noting that the long interbirth interval will itself contribute to the variance in the data since the difference between managing to produce one and two offspring during the sample period will be quite fine.

From these analyses, I conclude that the relationship between rank and fecundity is both robust and biologically real. Table 5.1 suggests that each loss of rank costs a female the equivalent of half an offspring over the 10-year reproductive lifespan that the average female can expect to have (Dunbar, 1980b). With the average lifetime reproductive output being only about five infants, this represents a loss of 10% of the female's potential contribution to the species' gene pool for each unit decline in rank. This is clearly a very significant selection pressure.

5.2.3 The physiological mechanism

Any explanation offering a functional interpretation for a particular phenomenon is inevitably given greater plausibility if it can be shown that a physiological mechanism that would produce the hypothesized effect actually does exist. By linking the two levels of explanation, we make it more difficult to reject either of them.

In the original analyses of the gelada data, I adopted a neutral strategy in so far as the likely mechanism was concerned in that I tried to let the animals tell me what the mechanism was. My approach was essentially to offer a series of alternative options and then test between them using relevant data from the field. There were two quite separate steps here: one was concerned with the physiological mechanism and the way it produces a loss of fecundity in low-ranking females; the other was concerned with the behavioural processes that give rise to this physiological effect.

The loss of fecundity can occur at three possible points in the sequence of events involved in reproduction, namely (1) prior to conception as a result of a failure by low-ranking females to mate with sufficient frequency to ensure that conception occurs, (2) at the point of conception itself, either because low-ranking females have a high frequency of anovulatory menstrual cycles or because the fertilized ovum fails to implant (or, alternatively, that the foetus is aborted during early pregnancy) and (3) post-natally due to low-ranking females suffering significantly higher frequencies of still-birth or infant mortality.

From the Sankaber data, we can demonstrate (1) that high-ranking females do not mate (or receive ejaculation) significantly more often than low-ranking females (t test, $p > 0.05$), (2) that when two females cycled together (i.e. were in oestrus during the same month), the higher ranking one was significantly more likely to conceive first (χ^2 test, $p < 0.05$) and (3) that neonatal mortality rates were too low to produce the documented loss of fecundity in low-ranking females (indeed, such mortality as there was tended to fall on the offspring of high-ranking females) (Dunbar, 1980a). Between them, these tests narrow down the options to the period around conception.

Reduced fecundability might be due to one of four causes: (1) a delayed return to oestrus following a period of post-partum amenorrhea, (2) a high frequency of anovulatory menstrual cycles, (3) a failure to implant following conception and (4) a high rate of spontaneous abortion during the first part of pregnancy. There is considerable evidence from experimental work on nonhuman primates as well as clinical and experimental studies of women to suggest that stress (both physical and psychological in origin) can cause temporary infertility (Bowman *et al.*, 1978; Grossman *et al.*, 1981; Wasser and Barash, 1983; Abbott, 1984; Yen and Lien, 1984). Stress, however, need not be the only explanation: body condition is known to affect fertility in many species of mammals (Sadleier, 1969), so that at least two alternative mechanisms can be considered as likely physiological explanations.

Testing these four options and their two underlying physiological mechanisms is rather difficult without the benefit of hormone sampling and energetic analysis. Nonetheless, we can test two predictions which between them can only be compatible with the stress-induced suppression of ovulation. Under this hypothesis, it should be true (1) that low-ranking females do not have longer periods of lactational amenorrhea than high-ranking females and (2) that low-ranking females do have longer oestrous[1] phases to their menstrual cycles than high-ranking ones. The first will be true if females are not prevented from returning to sexual receptivity by poor physical condition, while the second is a direct consequence of the suppression through stress of the luteinizing hormone surge that triggers ovulation (this being the normal event that switches off oestrus). The converse should be true if access to food is the factor limiting the reproductive rates of low-ranking females. There should be no differences between high and low ranking females on either of these measures if the poor performance of low-ranking females is due to either failure to implant or early abortion.

In the original analyses (Dunbar, 1980a), I tested the first of these predictions by plotting the age of the current infant at the time the mother came back into oestrus for the first time postpartum against the mother's dominance rank. Unfortunately, I also included in this sample a set of females whose first postpartum oestrus was not observed, but whose previous infant's ages

---

1. I use the term oestrus to refer to that period during which mating activity is particularly intense; in many species of primates (and the gelada in particular), this corresponds to the period when ovulation is most likely to occur and which is usually signalled by distinctive olfactory and/ or visual cues of receptivity (in the case of the gelada, by fluid-filled vesicles surrounding the areas of bare skin on chest and perineum: Dunbar and Dunbar 1974).

when the next infant was born could be estimated: for these, I estimated the age of the infant at first oestrus by subtracting the duration of gestation (six months) from its age when its younger sibling was born. But, this, of course, only estimates its age at the time its sibling was conceived, not the time at which the mother first started to cycle again. If we consider only the sample of females whose first postpartum menstrual cycles were actually observed, we find no relationship at all between the timing of the first postpartum oestrous period and the female's rank (linear regression: $y = 1.702 + 0.008x$, $r^2 < 0.001$, $t_{16} = 0.108$, n.s). The slope, though positive, is far too shallow to produce the loss of fecundity observed in low-ranking females. I tested the second prediction directly by comparing the lengths of the oestrus phases of the menstrual cycles of high and low-ranking females. Low-ranking females had significantly longer oestrous phases than high-ranking females, as predicted by the stress hypothesis (means of 14.3 and 10.7 days, $n=4$ and $n=3$ females, respectively; Mann-Whitney test, $p=0.05$).

One final set of observations lends further support to stress-induced reproductive suppression as the operative process. We can show (a) that the amount of harassment received by a female increases linearly with declining rank (Dunbar, 1984, figure 17: $r_s=0.720$, $n=30$, $p<0.001$) and (b) that, within females, the rate of harassment increases significantly when the female comes into oestrus (Dunbar, 1980a, table 7: Wilcoxon test, $n=10$, $p=0.05$). On average, the rate of harassment during the oestrous phase of the menstrual cycle (i.e. around the time when ovulation is likely to occur) was more than three times that received during the opposite period of the cycle.

Thus, there is strong prima facie support for the hypothesis that the poor reproductive performance of low-ranking females is caused by stress-induced suppression of ovulation. The case for the main alternative hypothesis (that low-ranking females are denied access to food) is clearly contradicted by the finding that the length of the lactational amenorrhea does not increase significantly with declining dominance rank. However, we cannot at this stage rule out the possibility that access to food acts indirectly via stress effects on fecundity. In other words, it may still be the case that fighting for access to food stresses low-ranking females and induces reproductive suppression indirectly. In the next section, I try to determine whether access to food plays any such role at all.

5.2.4 Stress versus resource access as the driving factor

Distinguishing unequivocally between these two hypotheses is more difficult than might appear at first sight because both depend on aggression as their *modus operandi*. Hence, any test of the access hypothesis is inevitably confounded by the stressful effects of the aggression used to displace competitors from a food source. Nonetheless, if food is of overriding importance, it should be possible to find situations in which the differentials are so great that the effects of stress *per se* are overridden. Such a situation is likely to occur if we consider the reproductive performance of whole units. The variance in resource quality within the area occupied by an entire herd of 200–400 animals (commonly an area 250 m or more in diameter) must inevitably be greater than that within the area occupied by a single reproductive unit (typically 15–35 m in diameter depending on the size of the herd and its location: Dunbar, 1986, table 15.3). Consequently, the advantages to be gained from defending resource patches are likely to be very much greater between different units than between individuals of the same unit. Hence, if these advantages exist, we ought to find a positive correlation between the size of a reproductive unit and its *per capita* reproductive rate (Wrangham, 1980; van Schaik, 1983). This is especially likely to be the case here because, in the gelada, females initiate all agonistic encounters between units, with the harem male becoming involved in only about 30% of all such encounters (Dunbar, 1983c): consequently, the number of females in a unit is likely to be the crucial factor affecting its ability to defend access to a resource patch. On the other hand, if stress is the most important factor independently of competition for food, then *per capita* reproductive output should correlate negatively with group size (again, the correlation should be strongest when plotted against the number of females in the group).

Table 5.3 gives the slope parameters for the regression equations of mean number of offspring per female on unit size for seven different bands at Sankaber and Gich. Overall, there is a fairly even distribution of positive and negative coefficients for mean number of offspring regressed on both the number of adults in the unit and the number of reproductive females. This speaks against the resource–access hypothesis. As a test of the stress hypothesis, however, this analysis suffers from a source of randomizing error that does not affect the resource–access hypothesis. This problem arises from the fact that there is considerable variance in the distribution of harem sizes among the different bands, as indicated by the figures for mean harem size in Table 5.3. Some bands contain a high proportion of very small units (i.e. units with fewer than three females). Such units are likely to be anomalous in their reproductive performance because the only way in which so small a unit can occur is if a large unit undergoes fission (Dunbar, 1984). Small units will thus tend to have the reproductive characteristics of the large units from which they derive. Although removal of the stressful effects of living in a large unit will result in the daughter units gaining a measurable improvement in reproductive rates, this gain will not filter through into as crude a measure as gross number of offspring until some years later. If we recalculate the regressions for the bands in Table 5.3 using only those units which had at least three reproductive females, we find that six of the seven bands have negative slope coefficients, as predicted by the stress hypothesis. Pooling the results using

Table 5.3

Slope parameters of equations for mean number of offspring per female in individual units regressed on total number of adults and subadults in unit, number of post-puberty females in unit, and number of females in units that had three or more females, for individual bands of the Sankaber and Gich populations

Band	Females/unit (mean)	All units			Females only			Units with > 2 females only				
		slope	r^2	n	slope	r^2	n	slope	r^2	n	$t\,(b=0)$	p^*
Main, Sankaber (1971)	3.48	0.003	0.00	25	−0.008	0.00	25	−0.045	0.03	17	−0.64	0.267
Abyss, Sankaber (1971)	4.67	−0.150	0.38	6	−0.207	0.68	6	−0.207	0.68	6	−2.94	0.021
Main, Sankaber (1974)	5.06	−0.098	0.01	18	−0.168	0.25	18	−0.082	0.09	15	−1.12	0.142
Abyss, Sankaber (1974)	3.70	−0.10	0.07	10	−0.191	0.19	10	−0.154	0.11	6	−0.87	0.216
E1, Sankaber (1974)	5.40	0.047	0.21	5	−0.050	0.11	5	−0.018	0.01	4	−0.15	0.448
E, Gich (1973)	4.13	0.081	0.12	8	0.115	0.22	8	0.149	0.29	7	1.43	0.893
K, Gich (1983)	3.06	0.036	0.00	18	0.024	0.00	18	−0.196	0.14	12	−1.30	0.112

*Probability that slope deviates from zero in a negative direction.

Fisher's procedure yields $\chi^2 = 23.546$ (d.f.$=2k=14$, $p=0.05$), indicating a significant and consistent trend.

The exceptional band in this case is itself interesting. It showed a very strong positive correlation between group size and mean reproductive rate no matter where group size was truncated. Significantly, it was also the only band that had a non-positive growth rate (annual growth rate of -0.3%: Ohsawa and Dunbar, 1984) and its ranging area was the highest in altitude of all the Simen bands. Birth rates are known to decline with altitude in the gelada (Ohsawa and Dunbar, 1984) and it is possible that as natural fecundity declines in the face of deteriorating habitat conditions, so access to food becomes an increasingly important determinant of fecundity such that it eventually overrides the stress-related costs of living in large groups. This highlights the point that we should beware of unitary explanations for phenomena: most biological processes are multifactorially determined and different factors may become prominent as environmental conditions change.

On balance, then, the evidence comes down rather firmly in favour of social stress as the cause of the poor reproductive performance of low-ranking gelada females. These analyses also suggest that access to resources is not the reason why gelada females live in groups, a result that we might have anticipated on the grounds that grasses (the staple diet of the gelada) are not a resource whose dispersion is normally so patchy as to make active defence a worthwhile proposition.

5.3 STRATEGIES FOR MINIMIZING INFERTILITY

Simulation modelling of the fitness consequences of forming coalitions with different individuals (Dunbar, 1984) suggests that female gelada rely on conditions formed with more dominant individuals to reduce the costs of living in large groups. The three main conclusions that emerged from these analyses were: (1) that, for demographic reasons, the most profitable coalition for a female is one formed between a mother and her daughter (especially the firstborn offspring), (2) that the selective advantage of a coalition seems to derive largely from a form of longterm reciprocal altruism in which the mother first supports her daughter while the daughter is a subadult and the daughter later pays back the 'debt' by supporting the mother when the mother is in her vulnerable old age and (3) that coalitions have no effect on the rank of the dominant member (and hence were assumed not to influence her reproductive performance).

There are two particular problems with these analyses. One is that it was assumed that the costs incurred by the dominant member of the coalition were negligible. The other was that the explanation in terms of reciprocal altruism was essentially an argument by default: kin selection was shown to make only a limited contribution to the female's inclusive fitness when compared to the advantages in terms of improved personal fitness. Yet, the fact that the 'debt' could only be repaid years later (and then at a time when the mother could exert no sanctions against daughters who reneged on their 'commitments') seems to tax even the advanced intellectual abilities of primates to their limits.

The costs of coalitionary support were ignored because they appeared to be negligible. On average, agonistic encounters occurred at a rate of only one every 5.9 hours per dyad, and of these only 5.5% (one every nine days) involved any kind of physical contact (Dunbar, 1980a). It is, of course, possible that the dominant members of a coalition might incur indirect costs in the form of stress when going to the aid of an ally. This particular cost, however, was by definition included in the estimates of rank-specific birth rates since these were obtained directly from the natural situation where all such costs must already have had their impact. Nonetheless, it is worth checking on this claim by asking whether the dominant member of a coalition incurs a loss of reproductive output compared to females of the same age and rank who are not members of a coalition.

Table 5.4 shows that the dominant member of a coalitionary dyad actually did better than females of the same age who were not members of a coalition. This result was rather unexpected, but it implies that even if dominant females do sustain a cost in terms of stress, they gain a personal advantage that more than offsets this cost. Moreover, they gain this reproductive advantage even though the coalition has no effect on their own rank within the unit. (In order to minimize the sources of error discussed above, I have here only counted the numbers of yearlings aged 6–18 months and the numbers of

Table 5.4

Estimated age-specific birth rates in relation to whether or not the female was a member of a coalition; for females who were in coalitions, a distinction is made between the dominant and the subordinate member

Female age class	Coalitionary status	Subordinate female		Dominant female	
		Birth rate	n	Birth rate	n
Subadult	Member	0.524	14		
	Non-member	0.500	4		
Young adult	Member	0.667	1	0.333	6
	Non-member	0.191	7		
Old adult	Member	0.533	5	0.417	8
	Non-member	0.285	7		

Note: birth rate per year estimated from the number of yearlings (aged 6–18 months) and the number of infants (aged 0–6 months) which each female had during 1974–5 in 11 units of the Main band at Sankaber.

infants aged 0–6 months that each female had.) Note, incidentally, that the subordinate partner also did better on average than a female of the same age who was not a member of an alliance: this confirms the claim that females form coalitions with high ranking females in order to gain a reproductive advantage.

These results raise an interesting question: if being in a coalition does not affect the rank of the dominant partner, how can it nonetheless affect her reproductive rate? Since we can find no evidence to implicate access to food as a factor, the only plausible explanation must be that she suffers less stress if she can count on at least some support (however limited). In other words, she may have to work less hard to maintain her position within the hierarchy and she may be challenged less often by other individuals. Table 5.5 confirms that the dominant female of a unit was involved in fewer agonistic interactions if she was a member of a coalition than if she was not. Note that this applies both to interactions with members of her own unit and to interactions with members of neighbouring units, though only in the latter case is the difference statistically significant (nonparametric t tests: $t_3 = 1.427$, $p < 0.15$, and $t_3 = 5.509$, $p < 0.01$, respectively, 1-tailed).

There is a second implication of the relationship between rank and fecundity that we can test. As units grow in size, the pressure on the females from stress-induced reproductive suppression will increase. Females in large groups should therefore want to invest more heavily in maintaining coalitions through time and this should result in a reduced interest in interacting with other members of the group. This is in fact the case. The number of other females that an adult female interacts with increases steadily with increasing unit size up to a maximum unit size of six females, after which it declines again (Dunbar, 1983a, figure 1). For the 43 females in units with 6–10 females, there is a significant negative relationship between number of female interactees and the number of females in the unit (linear regression: slope $b = -0.451$, $r^2 = 0.223$, $t_{41} = -3.427$, $p < 0.01$). This is confirmed by an analy-

Table 5.5
Mean frequency per hour with which the dominant female of a unit was involved in agonistic encounters

Coalitionary status	Mean encounter rate/hour		Females sampled
	Within unit	*With other units*	
Member	0.0098	0.8117	3
Non-member	0.0642	1.206	2

Source: focal samples on five reproductive units of the Main band, Sankaber, during 1974–5.

sis of the evenness with which females distributed their social time among the other adult members of their units: as size increases, so the evenness of their social interactions (as measured by Shannon's index of heterogeneity adjusted for group size) declines (Dunbar, 1984, figure 37: linear regression, $r^2=0.607$, $t_{23}=-5.964$, $p<0.001$). Figure 5.1 shows that this withdrawal from interaction is related to a tendency for females in large units to concentrate increasingly on their main social partners (i.e. their primary allies): the proportion of a female's social time that is devoted to her preferred social partner increases as the number of females in the group gets larger (linear regression: $r^2=0.411$, $t_6=2.048$, $p<0.05$, 1-tailed).

This social retrenchment by females correlates with a similar withdrawal from interactions with the harem male. The number of females that the unit's male groomed with at all increased in direct proportion to the number of females in the unit up to a unit size of five females, at which point it reached an asymptote (Dunbar, 1984). This asymptotic relationship correlates well with the risk that a unit runs of being taken over by another male (Dunbar, 1984). However, it was never clear to me why this should be so since I could see no good reason, given our understanding of how gelada society functions,

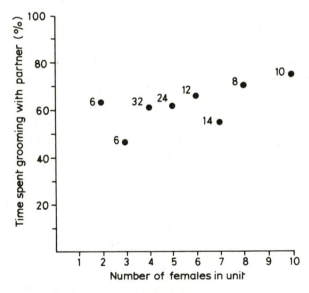

Figure 5.1 The mean proportion of a female's social interactions with the adult members of her unit that were with her main grooming partner, plotted against the number of reproductive females in the unit. The number of females sampled in each case is indicated by the number against each data point. The data derive from scan samples of units taken at 2-min intervals during social periods, recording who was interacting with whom (Dunbar, 1983a). Source: 11 units sampled at Sankaber, Simen, during 1971–2 and 13 units sampled during 1974–5.

as to why the amount of grooming that a female does with her male should be related either to her reproductive performance (the male's partner female aside: section 5.2.2) or to her loyalty to the male. Given the present results, however, the causal logic seems clear: the increasing pressure on females as unit size increases forces them to concentrate on reinforcing their critical relationships (i.e. those with coalition partners) and the constraints imposed by a limited time budget consequently oblige them to withdraw from inter-actions with less important members of the group. These include not only other females, but also the male (except, of course, in the case of the male's partner female, for whom the male is essentially a substitute female). This leads both to social fragmention and, because of reduced familiarity, to an increasing instability in the dominance relationships among the females (Dunbar, 1984). Thus, the male's failure to groom with all his females in large units is less a consequence of any decision on his part than a consequence of the way in which the females' behaviour changes. Similar time budget constraints force lactating mothers to withdraw from more casual interactions with other group members so as to concentrate on their key social partners as their time budgets are squeezed by the need to spend increasing amounts of time feeding in order to fuel the infant's growing demand for milk (Dunbar and Dunbar, 1988).

It is surely significant in this context that the crisis point at which both the females begin to reduce the size of their social networks and the unit becomes susceptible to takeover appears to occur at just the harem size where the unit first consists of two distinct nuclear families (Dunbar, 1983a). This is the point at which conflicts of interest over loyalties to collateral relatives and to a female's own mature daughters are first likely to occur. The simulation analyses have suggested that one reason why coalition formation functions so smoothly in this population of gelada is that the life expectancies of the females are such that a female is rarely likely to face conflicts between loyalty to her mother and loyalty to her eldest daughter when this individual matures into the adult cohort (Dunbar, 1984). Collateral relatives, on the other hand, are not so conveniently spaced in terms of relative age, so that conflicts of loyalty are more likely to occur between them.

5.4 CONCLUSIONS

I have tried to set out the various steps in the analysis of gelada female repro-ductive strategies as clearly as I can. I hope that I have been able to demon-strate two things: (1) that low-ranking females do suffer from a loss of fecundity and (2) that this loss of fecundity can only be explained in terms of the stress incurred from being harassed by higher ranking females. I have tried to demonstrate this (a) by presenting direct tests of hypotheses at each stage, (b) by disproving the possible influence of confounding variables and

(c) by undertaking direct tests between alternative explanations. To this battery of tests I would add a fourth consideration, namely the coherence of my explanation. Any attempt to discount either of the two key hypotheses has to be able to explain why there is evidence not only for the other hypothesis, but also for a series of subsidiary effects. Thus, doubts about the relationship between reproductive success and dominance rank not only to have to account for the fact that five independent tests of the hypothesis yield very similar results, but also have to explain how it is that we can have prima facie evidence for reproductive suppression without there being any detectable effect on reproductive output.

We can extend the logic of this argument for coherence one step further by pointing out the whole structure of gelada society can be explained as being a consequence of this one fundamental physiological effect (Dunbar, 1984, 1986). In other words, the way in which gelada society functions can be interpreted as a set of responses to the ramifying effects of this one phenomenon as its consequences spread in ever-widening circles through the species' socioecological system. Since we can provide direct empirical evidence for many of the claims arising out of this (e.g. fission rates, the distribution of unit sizes, the rates at which males take over units, the relationships between a male and his females and between females), this very coherence necessarily provides support for the claim on which they are all founded (namely, that female reproductive rates decline as a result of stress-induced reproductive suppression).

I have also attempted a direct test between the two main hypotheses that can be advanced to explain why gelada families live in groups. I have shown that the data on reproductive output are incompatible with the predictions of the hypothesis that most primates live in groups in order to defend access to food resources. The negative relationship between reproductive output and group size can only be explained on the assumption that females form groups in order to buffer themselves against the costs of harassment incurred when foraging in large herds of 200–400 animals. (The reasons why such large herds are formed need not detain us here; they are discussed in more detail in Dunbar, 1986.)

Finally, I have shown that, contrary to my earlier supposition, dominant females do gain improved fecundity from forming coalitions with lower ranking individuals. I was misled in this respect because I had assumed that coalitions could only influence a female's fecundity through their effect on dominance rank. The finding that dominant females do gain an advantage is important because it greatly strengthens the claim that coalitions have evolved through reciprocal altruism rather than through kin selection. We can now argue that there is a direct reciprocation of benefits rather than having to rely solely on long-term reciprocation (even though this still remains a significant contribution to the advantages of coalition formation). As is clear from

Table 5.4, the reproductive benefit of being a member of a coalition is quite considerable for the dominant female, though it is not as great as that obtained by the subordinate member of the coalition.

It is important to note, in conclusion, that these analyses do not demonstrate (and were never intended to suggest) that dominant females necessarily contribute more offspring to the next generation than low-ranking females do. The documented relationship between rank and fecundity relates solely to instantaneous reproductive rates (i.e. birth rates). If all other things were equal, then these differentials in fecundity undoubtedly would translate into comparable differentials in genetic fitness. But one of the hallmarks of primate social biology is their ability to use sophisticated social strategies to circumvent the disadvantages which are imposed on them by other aspects of their biology. Coalitions allow females to offset the reproductive disadvantages of low rank in such a way as to minimize the impact on lifetime reproductive output of marked differentials in instantaneous reproductive rates. Indeed, there is empirical evidence from the gelada to suggest that females do manage to equilibrate lifetime reproductive output despite the rank-dependent loss of fecundity that occurs in old age (Dunbar, 1984). This does not, however, mean that there is no selection for those physical and psychological traits that allow animals to achieve high dominance rank. The intensity of the selection pressure is obscured by the animals' abilities to capitalize on alternative ways of offsetting the costs of low rank. But any individual who tried to opt out of this arena altogether would perform so poorly that it would be heavily selected against.

ACKNOWLEDGEMENTS

The field work was supported by grants from the Science and Engineering Research Council (UK) and the Wenner Gren Foundation for Anthropological Research. I am indebted to those colleagues who have drawn my attention to ambiguities in the original analyses of female reproductive strategies, especially Joan Silk and Barb Smuts. I am also particularly grateful to Sandy Harcourt for his comments on the manuscript.

REFERENCES

Abbott, D.H. (1984) Behavioural and physiological suppression of fertility in subordinate marmoset monkeys. *Am. J. Primatol.*, **6**, 169–86.
Bowman, L.A., Dilley, S.R. and Keverne, E.B. (1978) Suppression of oestrogen-induced LH surges by social subordination in talapoin monkeys. *Nature, Lond.*, **275**, 56–8.
Dunbar, R.I.M. (1979) Structure of gelada baboon reproductive units. I. Stability of social relationships. *Behaviour*, **69**, 72–87.

Dunbar, R.I.M. (1980a) Determinants and evolutionary consequences of dominance among female gelada baboons. *Behav. Ecol. Sociobiol.*, **7**, 253–65.

Dunbar, R.I.M. (1980b) Demographic and life history variables of a population of gelada baboons (*Theropithecus gelada*). *J. Anim. Ecol.*, **49**, 485–506.

Dunbar, R.I.M. (1983a) Structure of gelada baboon reproductive units. II. Social relationships between reproductive females. *Anim. Behav.*, **31**, 556–64.

Dunbar, R.I.M. (1983b) Structure of gelada baboon reproductive units. III. The male's relationships with his females. *Anim. Behav.*, **31**, 365–75.

Dunbar, R.I.M. (1983c) Structure at gelada baboon reproductive units. IV. Integration at group level. *Z. Tierpsychol.*, **63**, 265–83.

Dunbar, R.I.M. (1984) *Reproductive Decisions: An Economic Analysis of Gelada Baboon Social Strategies*, Princeton University Press, Princeton, N.J.

Dunbar, R.I.M. (1986) The social ecology of gelada baboons. In: *Ecological Aspects of Social Evolution*, (eds. D.I. Rubenstein and R.W. Wrangham), Princeton University Press, Princeton, N.J., pp. 332–51.

Dunbar, R.I.M. and Dunbar, P. (1974) The reproductive cycle of the gelada baboon. *Anim. Behav.*, **22**, 203–10.

Dunbar, R.I.M. and Dunbar, P. (1988) Maternal time budgets of gelada baboons. *Anim. Behav.*, **36**, 970–80.

Grossman, A., Moult, P., Gailand, R., Delitaza, G., Toff, W., Rees, L. and Besser, G. (1981) The opioid control of LH and FSH release: effects of met-enkephalin analogue and naxolone. *Clin. Endocrinol.*, **14**, 41–7.

Kawai, M. (ed) (1979) *Ecological and Sociological Studies of Gelada Baboons*, Karger, Basel.

Kawai, M., Dunbar, R., Ohsawa, H. and Mori, U. (1983) Social organisation of gelada baboons; social units and definitions. *Primates*, **24**, 1–13.

Ohsawa, H. and Dunbar, R.I.M. (1984) Variations in the demographic structure and dynamics of gelada baboon populations. *Behav. Ecol. Sociobiol.*, **15**, 231–40.

Sadleier, R.F.M.S. (1969) *The Ecology of Reproduction in Wild and Domestic Mammals.*, Methuen, London.

Sokal, R.R. and Rolf, F.J. (1969) *Biometry*, Freeman, San Francisco.

van Schaik, C.P. (1983) Why are diurnal primates living in groups? *Behaviour*, **87**, 120–44.

Wasser, K. and Barash, D.P. (1983) Reproductive suppression among female mammals: implications for biomedicine and sexual selection theory. *Q. Rev. Biol.*, **58**, 513–538.

Wrangham, R.W. (1980) An ecological model of female-bonded primate groups. *Behaviour*, **75**, 262–300.

Yen, S.S.C. and Lien, A. (1984) Mammals: man. In: *Marshall's Physiology of Reproduction* (ed. G.E. Lemming), 4th edn, Churchill-Livingstone, London, pp. 713–88.

CHAPTER SIX

Reproductive strategies of subadult Barbary macaque males at Affenberg Salem

Jutta Kuester and Andreas Paul

6.1 INTRODUCTION

Intelligent decisions and highly developed social skills are necessary to cope with the complexity of social networks within primate groups. Many studies show that primates not only know their own social position in a group, but also the relationships between other group members and what can be anticipated in social interactions (Cheney and Seyfarth, 1980; Kummer, 1982; Dasser, 1986). The long-lasting nature of many social relationships – individuals may stay together for life – gives rise to the development of reciprocating social acts, especially reciprocal altruism, e.g. alliance formation among unrelated individuals. With respect to male mating strategies, several variables supplement individual physical strength during competition over females under these complex social conditions. Male 'power' can be regarded as a summation of intelligence, physical strength, age, temperament, experience, special social relationships, etc. Since most of these variables will change during the course of an individual's life, not only male 'power' itself but also the way in which it is expressed will change. It can therefore be expected that competitive strategies and, ultimately, male mating strategies will vary during the lifetime of the individual. In multi-male–multi-female groups, a male is embedded in a network of relationships with other males, all with their own specific pattern of power. An individual's mating strategy cannot therefore be viewed separately from those of other males.

Although it was recently realized that an analysis of distinct mating strategies is important for an understanding of how males can maximize their reproductive success (Robinson, 1982; Berenstain and Wade, 1983), most studies attempted to evaluate mating success in macaques and baboons on the basis of dyadic rank (Fedigan, 1983 gives details of methodological problems involved with this topic). Studies frequently inferred male reproductive success from consorting activity which, however, is only one reproductive strategy, although probably a very successful one. No longitudinal studies have been performed and these are essential for estimating lifetime reproductive success in males.

This chapter presents data on mating success and mating strategies of subadult Barbary macaque males at the beginning of their reproductive careers. Barbary macaques show a moderate sexual dimorphism with male weight exceeding that of females by about one third. The Barbary macaques at Salem are organized into multi-male–multi-female groups as in the wild. The sex ratio of mature individuals is almost equal (Salem 1:1.3, feral groups 1:0.9–1:1.2; Paul and Kuester, 1988). Barbary macaques show a pronounced reproductive seasonality with matings restricted to autumn and winter (MacRoberts and MacRoberts, 1966; Roberts, 1978; Kuester and Paul, 1984). Females exhibit distinct estrous periods with corresponding changes in anogenital swelling and mating activity, these enabling a relatively accurate estimation of the fertile period of a cycle. More than 90% of all conceptions at Salem occurred during the first estrus of a mating season (between October and December). Almost all females show one postconceptional estrous period (Kuester and Paul, 1984).

In contrast to the two other most intensely studied macaque species (*M. mulatta, M. fuscata*), the Barbary macaque is a single mount-to-ejaculation species. An ejaculatory copulation can be achieved within a few seconds. It will be shown that this mating pattern has important consequences for male mating success.

6.2 MATERIALS AND METHODS

Data for this study were collected during two mating seasons (1984–85, 1985–86) in two groups (B and C) of the Barbary macaque population at Affenberg Salem (Federal Republic of Germany). Affenberg Salem is a 14.5 ha area of mixed beech and spruce forest enclosed by an electrical fence. In addition to a variety of natural food sources utilized by the animals, they are also fed daily with grain (wheat or oats), fruit or vegetables. Water and commercial monkey chow are available *ad libitum* (Kaumanns, 1978; de Turckheim and Merz, 1984; Paul and Kuester, 1988).

The study subjects were twelve males born in 1980, i.e. they had just reached sexual maturity at the beginning of the study. Seven males were still living in their natal group (group B), five were non-natal members of group C. They transferred to group C one year before the start of the study. Four of them were born in group B, the fifth in a third group of the population. When the males migrated into group C, the only natal male of their age left. He became a peripheral member of group B and was not included in the study. In both groups, matrilineal kinship relations of all but the oldest, wild-born individuals (i.e. born before 1971) are known. The social history of all study subjects (rank and parity of the mother, age and sex of siblings, social relationships) is known in detail from birth. The age–sex composition of the groups (sexually mature individuals only) for the two observation years is given in Table 6.1.

Data were collected with focal animal sampling and *ad libitum* sampling techniques. A focal observation session lasted one hour. Each male was observed for 5–8 h/month during the mating season 1984–85 and 10 h/month during the mating season 1985–86. Focal animal sampling was discontinued each year after December when sexual activities almost ceased (Kuester and Paul, 1984). During observation of a focal male, all matings, grooming, agonistic interactions, and consort activities of other group members which occurred in the vicinity of the focal male were recorded (*ad libitum* data). This combined form of data collection was used by the authors for several years. A comparison of mating frequencies of individual males during focal animal and *ad libitum* observations showed significant to highly significant positive correlations, so that these *ad libitum* data could be regarded as representative of an individual's mating success.

Each female was checked daily for her reproductive state. Based on the birth dates of the subsequently born infants, an estrous period could be classified either as a *conceptional estrus* (CE), *postconceptional estrus* (PCE) or *nonconceptional estrus* (NE). Hormonal conditions during the first two types of estrus could be defined (ovulatory cycle leading to a successful pregnancy and

Table 6.1

Age-sex composition of sexually mature group members. Study subjects underlined: ( ) natal males; *postmenopausal, sexually inactive females excluded; **one female transferred from group B to Group C during the mating season

Season		1984–85		1985–86	
Group		B	C	B	C
Males (age)					
Adult	(7+)	22(9)	8(0)	24(12)	10(0)
Subadult	(6)	3(3)	2(0)	3(3)	1(0)
	(5)	3(3)	1(0)	8(7)	5(0)
	(4)	8(7)	5(0)	9(9)	7(6)
Females	(3+)*	62	27	72.5**	28.5**
Sex ratio					
Adult		1:2.8	1:3.4	1:3.0	1:2.9
Mature		1:1.7	1:1.7	1:1.6	1:1.2

cycle during early pregnancy, respectively). Nonconceptional estrous periods were heterogeneous. They included anovulatory cycles and ovulatory cycles which did not lead to a successful pregnancy owing to lack of implantation or early abortion. All types of estrus ended abruptly within 24 h. Sexual activities ceased completely after 1–4 weeks of continuous male–female associations (consort relationships) and a high mating frequency. The onset of estrus was more vaguely defined since sexual activity and attractivity increased gradually. For data analysis estrous periods were divided into 7-day intervals (counting backwards from the day of cessation of mating). In the following sections, a female is termed 'highly attractive' or 'at peak estrus' during the last seven days of an estrous period. Matings which terminated with an ejaculation are termed complete or successful matings.

6.3 RESULTS

6.3.1 Mating success of subadult males

For 1984–85, total mating frequency, as well as the number of complete matings, of the 4-year old study subjects were considerably lower than they were one year later (Table 6.2). In each group, all but one of the males showed an increase in mating and ejaculation frequency. Mating frequency of the non-natal C group males was lower than that of the natal B group males for both years. Owing to the high degree of individual variation present, this difference was not significant (Mann Whitney U-test; $p > 0.05$). The frequency of complete matings for B group males during the second year tended to be higher than that for C group males ($p < 0.1$). When the fact that group B had more than twice as many mature females as group C in both years is taken into account, 'access' of C group males to females, i.e. mating frequency per mature female, was, however, higher than that of B group males. Similar focal data on adult males are not available to date so that a direct comparison of mating frequency between subadult and adult males is not possible. *Ad libitum* observations indicate that subadult males have a higher mating frequency than young adult males (7–9 years of age) but lower than most older adult males. Ejaculation rate of the focal males could be compared with data collected in group B during 1982–83 on females during all stages of their cycles. Adult males (7 years and older) had an ejaculation rate of 40% ($n = 207$), this being higher than that of the study subjects except in group C during the first year (Table 6.2). Again, due to high individual variations, the difference in ejaculation rate between the groups was not significant (Mann Whitney U-test; $p > 0.05$).

Rate of ejaculation depends largely on female cycle stage with the highest values during the peak of estrus (unpublished data). An accurate estimation of the sexual performance of subadult males must, therefore, be based on a

Table 6.2
Mating frequency (n/h) and ejaculation rate (%) of the focal males

Season	1984–85		1985–86	
Group	B	C	B	C
Total matings	0.65	0.30	1.00	0.66
Range	0.32–1.40	0.05–0.63	0.50–1.58	0.30–1.10
Complete matings	0.19	0.13	0.35	0.24
Range	0.05–0.41	0–0.26	0.17–0.53	0.13–0.30
Ejaculation rate	29.3	44.8	34.9	36.4
Range	9.7–56.3	0–100	28.6–60.0	27.3–50.0

Table 6.3
Relative mating frequency (%) during different stages of the female cycle (definition of cycle stages see below)

Season	1984–85				1985–86			
Group	B	C	B	C	B	C	B	C
Data	Focus		Ad lib.		Focus		Ad lib.	
Cycle stage								
EW	28.3	37.9	40.5	45.2	30.1	24.2	37.8	34.8
EW−1	31.3	31.0	19.2	16.8	21.6	26.3	14.5	19.3
PCW	7.1	3.5	15.0	13.5	18.2	13.1	19.2	21.2
Rest	33.3	27.6	25.3	24.5	30.2	36.4	28.6	24.7
n	99	29	818	155	209	99	871	598

comparison of matings during identical stages of the cycle. Table 6.3 shows the distribution of matings during four different cycle stages: (a) EW: the last week of CE or NE, (b) EW−1: the week before EW, (c) PCW: the last week of PCE, and (d) all remaining matings, most of which occurred before EW−1 of the first estrus of a season. Data on focal subjects were compared with *ad libitum* observations where males of all ages were included. *Ad libitum* data of both groups during both study years showed almost identical distributions with the highest values during EW and lowest values during EW−1 and PCW, respectively. The young males had a different mating pattern. Highest values were found during the 'rest' stage followed by EW−1, i.e. they mated primarily with females not fully estrous and, therefore, not very attractive. This difference was significant in group B in both years and in group C in the second study year (Chi square-test; $p < 0.05$; comparison of mating frequency during EW versus all other matings). The high values for C group males during EW in 1984–85 reflect their very low mating frequency, a single copulation carrying a disproportionately high weight in a presentation based on relative values. Males from group B increased their mating frequency with highly estrous females slightly during the second study year, while C group males showed a considerable decrease.

Mating frequency and access to females at the peak of conceptional estrus, when ovulation occurs, is crucial in estimating a male's reproductive success. Focal males of both groups in both years were able to mate with females during that period (Table 6.4). Mating frequency increased from the first to the second study year in both groups, but that of C group males was always lower than that of the natal males in group B. Ejaculation rate showed marked group differences. Males of group C had a high rate of successful matings, this remaining constant, while sexual performance of B group males was considerably lower at the beginning but increased from four to five years of age (Table 6.4). Consequently, the focal males in group C had a higher frequency of successful matings in both years (1984–85: 1.2 vs. 0.7/male; 1985–86: 3.0 vs. 2.4/male). Sexual performance of subadult males was low in both groups compared with adult males, whose ejaculation rate during CW was 86% (unpublished data from group B, 1983–84). Comparing mating success for subadult males during CW with the number of conceptions showed that their probability of siring an offspring appeared to be minimal. It should, however, be borne in mind that focal observations covered only a small proportion of the males' total activities (maximum of 10 h/male/month). These data compare well with those from a study during the mating season 1983–84 on group B (unpublished data). Nineteen focal females of all ages were observed during their last week of CE for a total of 121.7 h. A total of 240 complete matings were recorded, only 6.6% by four and five year old

Table 6.4

Mating frequency per male and ejaculation rate (%) of focal males during last week of conceptional estrus (CE)

Season	1984–85		1985–86	
Group	B	C	B	C
Frequency	3.1	1.6	5.7	4.0
Range	0–5	0–3	2–11	2–9
Ejaculation rate	22.7	75.0	42.5	75.0
Number of conceptions	48	22	42	26

males, who, however, represented 16.2% of the mature male class during that year.

6.3.2 How to achieve copulations: mating strategies

For an understanding of the mating strategies in this species as well as for estimates of a male's lifetime reproductive success, it is important to determine how these young males obtained access to such 'valuable' females. All focal males were clearly subordinate and physically inferior to the adult males, which heavily controlled all contacts with females at peak estrus. Moreover, even at the height of a mating season the number of adult males in both groups was much higher than that of sexually receptive (attractive) females. Two typical examples from focal observations will illustrate the reproductive strategies of four and five-year old males:

(a) Group B, 5 November 1983, actors: ♀43 (focal animal, about 14 years old, 4th day of CW), ♂227 (7 years, ♂U7 (5 years).

At 10:53, ♀43 transferred from her former sexual partner, who started consorting with another female, to ♂227. They mated successfully three times (10:54, 11:30, 11:43), groomed and huddled between the matings and remained undisturbed at the group's periphery the whole time. At 11:46, ♂U7 approached. He remained at 5–10 m distance, sunning, feeding, and grooming himself, while ♂227 and ♀43 huddled and groomed. No interactions occurred between ♂U7 and the pair. Fifteen minutes later, at 12:01, an adult male from a neighbouring group passed at a distance of about 20 m. Male ♂227 left the female and chased the strange male away. The moment he left, ♂U7 rushed towards the female who presented, ♂U7 mounted and started thrusting. When ♀43 started the mating call (after about two pelvic thrusts), ♂227 turned on his heels, racing towards ♂U7 and ♀43. However, ♂U7 completed the copulation with an ejaculation after a total of seven pelvic thrusts before ♂227 reached the scene and chased ♂U7 away. At 12:02, ♂227 returned to ♀43 and groomed her. Male ♂U7 also returned and stayed at 10 m distance for a total of 40 min before he eventually left. Animals ♂227 and ♀43 remained together grooming and huddling until the end of observation session (12:49). No matings occurred during this time.

(b) Group B, 10 December 1985, actors: ♂X20 (focal animal, 5 years), ♀U7 (7 years, 3rd day of CW), ♂227 (9 years), ♂152, ♂157, ♂159 (all 13 years).

At 11:31, 1.5 min after start of observation session, ♂X20 approached the pair ♂227/♀U7 and stayed at a distance of about 10 m. The pair fed close to one another (within arm's reach). They had mated successfully at 11:25 (*ad libitum* observation), but it was unknown how long they had already been together that morning. At 11:32, ♀U7 left, ♂227 followed her, and ♂X20 in turn followed the pair at a distance of about 20 m. After approximately 50 m, at 11:35, ♂227 stopped when ♂157 and ♂159 became visible (at 15 m and 20 m distance from ♂227 respectively). Male ♂X20 passed ♂227 and still followed ♀U7 who eventually stopped almost equidistant from all three adult males. ♂X20 inspected her genital area twice and they then started feeding, while the three adult males remained at their positions, sitting, yawning (indicating tension), and peering secretly at each other and the female. While feeding the female (and also ♂X20) gradually shifted towards ♂159 and were about 10 m from him, when, at 11:37, a fourth adult male, ♂152 appeared on the scene and mounted ♂227 (an interaction regarded as a means of reducing tension). At the moment of mounting ♂X20 mated with ♀U7 (ejaculation occurred after eight pelvic thrusts). Probably attracted by the female's mating call, ♂X20 was chased away (by ♂152). The other males had left. ♂152 returned and approached ♀U7. ♂X20 also returned and followed the pair. At 11:39, 2 min after ♂X20, ♂152 mated successfully with ♀U7 despite severe disturbance by ♂X20 who approached to 2 m distance and screamed. He was threatened by ♂152 (open-mouth threat), and retreated to a distance of 5 m, while ♂152 started grooming the female. ♂X20 left 1.5 min later and did not return to ♀U7 until the end of the observation session (12:30). How long ♂152 and ♀U7 remained together was unknown.

In both examples, the subadult males mated opportunistically. They 'sneaked' copulations during a moment when the consort male was distracted or during a situation where several adult males 'obstructed' one another, owing to their similar competitive power, dyadic dominance of one male (♂227) being outbalanced by the risk of a coalition between his rivals (♂152, ♀157, ♀159). The latter 'sneaking' situation was more common. It was extremely rare for females to hesitate in mating with young males at such moments.

Sneaking a copulation in full view of adult males was restricted to four and five-year old males (unpublished data from *ad libitum* observations). Five-year old males had better chances than 4-year old males, probably because of their higher dominance rank. Opportunistic mating was also observed in adult males of low social status, but only if the situation was 'safe', e.g. when males started fighting over access to the female and could therefore no longer control her contacts. The chances of these males were, however, low because they were peripheralized during the mating season.

Sneaking copulations was not the only mating strategy adopted by subadult males. Like adult males, they established consortships. These, however, were restricted to females not fully estrous, in which the adult males were not very interested and, hence, such associations were tolerated. The lack of attractivity was also reflected in the behaviour of the subadult males.

100

The consortships were often of short duration and also frequently spontaneously teminated by the males themselves. Such behaviour was only shown by males with attractive partners when they could transfer to an even more attractive female. Subadult males had no chance to establish long-lasting sexual relationships with females at peak estrus.

If mating frequency of the focal males during the last week of CE is considered (Table 6.4), it becomes clear that sneaking opportunities were rare, despite a high rate of consort exchange. Mean interval between successful copulations with different males was only 32.2 min in the study on peak oestrous females in 1983–84, and was probably very similar during the present study. The focal males frequently followed a single female and her partner(s) for a complete observation session (60 min) without obtaining (or taking) an opportunity to mate.

6.3.3 How to assure mating success: Harassment as a reproductive strategy

A behaviour exclusively shown by subadult males – especially five-year old subadults – was harassment of dominant males during matings as described in the second sample. In half of all harassments ($n=38$; both groups and years combined) the young male approached the pair rapidly and emitted various forms of vocalization, usually screams sometimes mixed with grunts. Other forms of harassment were silent approach ($n=14$) or screaming without approach ($n=21$). Approach always stopped prior to body contact with the pair. Touching or hitting (Niemeyer and Anderson, 1983, give details of this type of harassment in other primate species) was never observed. Harassment was performed in 1984–85 by all focal males, in 1985–86 by all but one male in group B. Although harassment refers to a behaviour of subordinate individuals, *ad libitum* data, where all ages were included, showed that it was a behaviour of *subadult* (very rarely juvenile) males directed towards *adult* males (Table 6.5). Although clear dyadic dominance relations also existed between many adult males, harassment was never observed among them. Disturbances of matings by females were extremely rare (group B: $n=11$, group C: $n=5$; both years combined).

A conspicuous behaviour such as harassment which, moreover, appeared to be 'useless' as suggested by the example given above, appeared worthy of detailed analysis. The main questions posed were: what are the costs for the actor in terms of receiving attacks from the target male or other group members? What are the costs for the recipient in terms of mating success? Is harassment related to the female's cycle stage? Is harassment related to the actor's own sexual relationships?

An analysis of the males' responses to harassment showed that the risk for the harasser was surprisingly low. The most common responses to the focal

Table 6.5
Age of actor (*A*) and recipient (*R*) of male harassment (*ad libitum* data, no harassment was observed in group C during 1984–85). * one case of simultaneous harassment of two males

Season	1984–85		1985–86			
Group	B		B		C	
Age	A	R	A	R	A	R
7+	0	23	0	50	0	21
6	4	0	1	3	0	2
5	5	0	37	2	13	2
4	14	1	15	0	13	1
3	0	0	3	0	0	0
?	1	0	0	0	0	0
Total	24	24	56*	55	26	26
% ad lib. copulations	2.9		6.3		4.3	

males for 73 recorded harassments during focal observations were ignoring (54.8%) or threatening (27.4%). Contact aggression occurred only once (in group B) when the harasser was chased and bitten by the dominant male and received a small wound on one foot. Other group members interfered in only 4.1% of all cases. They either chased the harasser, the female, or started a quarrel with the dominant male. Responses to harassment did not change with the age of the focal males. Males of group C, however, tended to be more intolerant of the harassers than males of group B. While 60% of all harassments in group B were ignored, only 33.3% were ignored in group C (χ^2=3.28, p < 0.01; ignoring was tested against all other responses directed towards the focal males). The most common response in group C was threatening (53.3%). A somewhat higher risk to group C harassers was also suggested by a relatively lower rate of harassment (2.0% vs. 3.4% of all *ad libitum* copulations, data for both years combined; χ^2 = 3.73, p <0.1).

In order to estimate the costs to the 'victims' of harassment, the ejaculation rate of disturbed matings was compared with that of undisturbed matings. Since ejaculation rate depended on cycle stage, undisturbed matings of the respective female on the same day and (to increase sample size) the day before and after were used for comparison. In the first year, harassment in group B significantly decreased ejaculation rate (χ^2 = 5.98, p < 0.02) and also tended to lower it in group C (Fisher test, p = 0.065). One year later, harassment of the focal males had no significant negative effect on ejaculation rate (p > 0.1).

Focal males of both groups during both years disturbed a higher percentage of matings with females at peak estrus than during other stages of the cycle. A significant or highly significant deviation from a chance distribution could only be verified for the second year (group B: $\chi^2 = 4.04$, $p < 0.05$; group C: $\chi^2 = 7.75$, $p < 0.01$. Matings during EW were compared with all other matings and only data from females whose matings were disturbed by the focal males were included). These results suggested that harassment was related to reproduction (not just mating). Lack of sexual experience should be considered for the weaker discrimination of the males at the age of four years.

If harassment was related to sexual competition between subadult and adult males, its timing appeared to be poor since the males disturbed ongoing matings, i.e. when the target male had already achieved intromission. In fact, subadult males did also harass sexual interactions occurring before mating. They approached and/or screamed during genital inspections and mounting attempts. Non-sexual interactions of a pair (e.g. grooming, huddling, feeding in close contact) and sexual behaviour of the female (e.g. presenting) were not disturbed. Harassment of matings and precopulatory behaviour were often repeated with the result that a consort exchange occurred before the first male had achieved a complete mating. The second male, however, was usually also disturbed. Sometimes the focal males stopped harassment and left the pair before a complete mating had occurred but they always departed if an ejaculation had occurred, usually immediately after the mating.

Table 6.6 shows the number of harassment episodes recorded during focal observations. An episode was defined as the period of continuous disturbance of sexual interactions (matings and 'premating' behaviour) of the same female. It could include several male 'victims'. For 4-year old males, most harassments occurred spontaneously. When they were five years old, however, most harassment episodes occurred when a focal male's own sexual contact with the female had been disrupted by the opponent (sexual contact included observed matings and assumed matings, i.e. when the focal male and the female were found close together at the start of an observation session). This age difference with respect to harassment occurrences was highly significant in group B and a similar trend was found in group C (Group B: $\chi^2 = 7.33$, $p < 0.01$; Group C: Fisher test, $p=0.13$). Males of group C tended to harass

Table 6.6
Harassment episodes (for definition see text) during focal observation

Season	1984–85		1985–86	
Group	B	C	B	C
Total n	26	5	54	15
After own sexual contact (%)	35.3	40.0	66.7	80.0

the male superseding them more frequently if sexual contact between the focal male and the female had taken place during EW (Fisher test, $p=0.08$).

Only one harassment episode (in group B) ended with a 'total success' for the focal male. Due to his harassment, he was able to re-establish contact with 'his' female before the adult male achieved ejaculation. Since the female was not at peak estrus on that day, it might have been easier for him to prevent the adult male from mating at this time than later. All other episodes, at best, ended with a delay of the next successful mating. In the following analysis, harassment was regarded as completely unsuccessful if the opponent achieved an ejaculation during the first disturbance of the focal male. Delay of success-ful episodes was calculated from the first disturbance of the focal male to either a complete copulation, the departure of the focal male or the end of the observation session. Exact intervals between the first harassment and a successful mating were often not known in the latter two cases, so that the values presented in Table 6.7 are minimum values. It was assumed, in this analysis, that a delay in mating was due entirely to the behaviour of the harasser and not influenced by other group members.

Harassment by B group males was more often successful than unsuccess-ful. In both years, their rate of success was higher after own sexual contact than during spontaneous harassment. Observations suggested that this was due to more intense harassment and also to frequent harassment of premating activities, the latter rarely occurring spontaneously. Not only rate of success but also delay of another male's mating was longer after own sexual contact. Maximum values show that the males spent considerable energy in preventing other males from mating. In some cases they did this successfully for more than one hour. Success rate of C group males was lower (not significant), and success of harassment after own sexual contact was even lower than during

Table 6.7
Success of harassment and delay of next mating *if* successful

Season	1984–85		1985–86	
Group	B	C	B	C
No success (%)				
all episodes	46.2	60.0	43.6	46.7
after own sexual contact	22.2	100.0	36.1	50.0
Mean delay (min)				
all episodes	13.0	11.0	8.0	9.0
after own sexual contact	23.7		9.5	9.5
Maximum value	67+	16+	36+	30+

spontaneous harassment. If successful, delay was similar in both groups. Differences in the success rate between the two study years were not significant.

6.4 DISCUSSION

'Sneaking' copulations with highly estrous females as an alternative mating strategy were clearly facilitated – perhaps only possible – under conditions of reduced female defensibility (Berenstain and Wade, 1983). Four factors seem to be primarily responsible for the comparably high chances of reproduction for young males: (1) Sex ratio of the groups showed a high number of mature males per female. This was even more accentuated with respect to the sexual context, because mature males always outnumbered estrous females ('effective' sex ratio $> > 1$). Males were sexually active (at least potentially) during the whole mating season, which lasted several months, while a female was sexually active and attractive for a total of only 4–6 weeks. (2) Reproductive seasonality reduces the benefits of monopolizing a single female, because a male will lose or at least greatly diminish his chances of fertilizing other females. Although exclusive consortships between adult males and females occurred in both groups, it was more common for several males to 'circulate' amongst the estrous females. These also showed a strong tendency to leave a male after a successful copulation. (3) Low 'power asymmetry' between males frequently led to a 'stalemate' which gave the subadults their chance. (4) Barbary macaques as 'single mounters' are able to complete a copulation literally within seconds. Low ranking males are therefore able to gain maximal profit from a distraction of a higher ranking opponent or from situations of low 'power asymmetry'.

Subadults sneaking copulations is of course disadvantageous from the viewpoint of adult males. Not all sneaking attempts are successful, however, because dominant males also interfered during matings and premating activities of subadults. Moreover, it is not known whether subadult males are as fertile as adult males. Japanese macaque males, which are sexually mature at the same age as Barbary macaques, do not reach adult testis size and testosterone levels before 6.5 years old (Nigi *et al.*, 1980). Ejaculation volume and sperm number were not analysed. Until further studies on this topic are performed, we must assume that all complete matings during the fertile period of a female have equal chances of fertilization. This is, however, probably not the case.

Sexual associations of young subadult males could easily be disrupted by adult males, simply by a close approach or a threat. This was in sharp contrast to the pattern of consort exchange between adult males. Their exchange was much more hesitant and clear-cut dominance behaviour was shown only rarely. This resulted in longer transition periods and longer mating intervals.

However, the subadults could frequently delay the next copulation after loss of their mates by harassment. The male 'victims' had only limited counter-strategies. Above all, they had to maintain contact with the female and not perform conspicuous actions which would attract other males. Under conditions of frequent consort exchange, any male strategy which delays the mating of his successor during a female's fertile period will increase his own chances of fertilization because long mating intervals decrease the number of competitors. Harassment is the means by which subadults achieve this goal. Apart from this, a delay may also be of advantage in giving the harasser's sperm a 'head start'. If this were true, harassment should stop spontaneously after a certain period of time. The present study could not prove this, perhaps because observation sessions lasted only one hour.

We have no effective explanation for spontaneous harassments. The considerable decrease in the second study year suggests ontogenetic effects. Spontaneous harassment was often directed towards males who mated with 'favourite' females of the harasser (indicated by frequent contact before the females reached peak estrus).

One male, the highest ranking of his peer group, no longer harassed at five years old. He was the only male of his age which already dominated several older males and had challenged many others. During the next meeting season, when the study subject were six years old, all but one male (lowest ranking of his age) had stopped harassment, and several of them had challenged older males, although mostly unsuccessfully. This suggests that harassment is only performed as long as clear-cut, undisputed dominance relations exist between the males.

Differences in sexual activities between natal and non-natal males were related to differences in group size, sex ratio, and familiarity with adult males. Non-natal males lived under more restrictive conditions. Group C had fewer females and, during the second year, a higher number of mature males per female. Relationships between the subadults and most adult males were tense (inside and outside the mating season). All these factors led to a lower mating frequency, more limited access to highly estrous females and a lower frequency of harassment with a slightly higher risk. However, if the subadult males of group C had a chance of mating they were more 'efficient' than the natal males. They had a higher rate of complete matings and a higher rate of harassment after own sexual contact, especially after contact with peak estrous females. Thus, less favourable social conditions may accelerate social maturation and the development of social skills. Mating strategies of natal and non-natal males showed no differences.

Comparative data on mating success and mating strategies of subadult males in other macaque species are scarce. Sexual activities of adolescent rhesus and Japanese monkeys are very restricted. They are often spatially separate from females during the mating season, spending their time at the group's periphery or in all-male bands (Conaway and Koford, 1964; Kaufmann, 1965; Stephenson, 1975; Takahata, 1980, 1982). Harassment of dominant males by subordinates is rare in these species (Niemeyer and Anderson, 1983. Van Noordwijk (1985) observed disturbances of matings by subadult males in wild long-tailed macaques. Reproductive chances for these males appeared to be nonexistent because exclusive consortships of the two highest ranking males during the probable ovulation period were observed in four out of five conceptions. A study on mating activities in *Macaca radiata* (Glick, 1980) also related mating success to female's cycle stage. Subadult males had access to females during the estimated conception period but their access was restricted to less attractive females (adolescent and low ranking females). Only adult males established long-lasting consort relationships. Groups with a clear spatial separation of subadults were also observed in this species (Simonds, 1973). Taub (1980) studied mating patterns in one wild group of Barbary macaques. Males who were estimated at four and five years of age respectively had a low mating frequency. Harassment and interventions were not observed and tolerance amongst males was emphasized. Taub described different male mating strategies and compared their relative mating success. Sneaking copulations, as was typical for the subadults in the present study, were not mentioned. Unfortunately, a determination of the type of estrus and cycle stage was not possible in Taub's study, and this may, to some extent, be responsible for the different findings.

The present study shows that focussing on individuals which may be regarded as less 'successful' or even 'unimportant' can elucidate alternative mating tactics and may explain important phenomena otherwise poorly understood or even not realized. This approach may be crucial for a better understanding of evolutionary traits in primates and in the analysis of how individuals cope with problems, elucidating the conditions which enable and also limit their actions. Further such studies which consider the dynamic aspect of relationships and interdependence of individuals within social groups are, however, necessary.

6.5 SUMMARY

Twelve subadult Barbary macaque males (natal and non-natal) living in two different social groups were observed during the first two mating seasons after reaching sexual maturity. Mating frequency and ejaculation rate increased from the first to the second study year but remained lower than in adult males. The study subjects had a disproportionately high rate of matings with females not fully estrous but matings with females at peak estrous also occurred, suggesting that they had some, although low, chances of fertilization. The subadults mated opportunistically with highly estrous females during moments of distraction of the consort male or in situations of low

107

power asymmetry among adult opponents which prevented them interfering. This mating strategy depended on frequent consort exchanges, low female defensibility and the single-mount-to-ejaculation mating pattern of this species. Due to their low dominance status subadult males could not establish long-lasting sexual relationships with peak estrous females, but they frequently delayed the mating of the next male(s) by harassment, a behaviour shown almost exclusively by this age class. An action like harassment which delays the mating of an opponent is regarded as a reproductive strategy which enhances the chances of fertilization for the actor either because an increased mating interval during the fertile period of a female decreases the number of competitors or competitors' sperm or because a longer mating interval gives the actor's sperm a better start. Group differences in frequency of mating, ejaculation and harassment indicated that the non-natal males lived under more restrictive social conditions, but they made better use of their opportunities than natal males.

ACKNOWLEDGEMENTS

We would like to thank Walter Angst, Ellen Merz, Gilbert de Turckheim and Christian Vogel for their support and encouragement. This research was financially supported by Deutsche Forschungsgemeinschaft (An 131 1–5).

REFERENCES

Berenstain, L. and Wade, T.D. (1983) Intrasexual selection and male mating strategies in baboons and macaques. *Int. J. Primatol.*, **4**, 201–35.

Cheney, D.L and Seyfarth, R.M. (1980) Vocal recognition in free-ranging vervet monkeys. *Anim. Behav.*, **28**, 362–67.

Conaway, C.H. and Koford, C.B. (1964) Estrous cycles and mating behavior in a free-ranging band of rhesus monkeys. *J. Mammal.*, **45**, 577–88.

Dasser, V. (1986) Social concepts of monkeys. *Primate Rep.*, **14**, 59. (Abstr.)

Fedigan, L.M. (1983) Dominance and reproductive success in primates. *Yearb. Phys. Anthrop.*, **26**, 91–129.

Glick, B. (1980) Ontogenetic and psychobiological aspects of the mating activities of male *Macaca radiata*. In: *The Macaques: Studies in Ecology, Behavior and Evolution* (ed. D.G. Lindburg), Van Nostrand Reinhold, New York, pp. 345–69.

Kaufmann, J.H. (1965) A three-year study of mating behavior in a freeranging band of rhesus monkeys. *Ecology*, **46**, 500–12.

Kaumanns, W. (1978) Berberaffen (*Macaca sylvana*) im Freigehege Salem. *Z. Kölner Zoo*, **21**, 57–66.

Kuester, J. and Paul, A. (1984) Female reproductive characteristics in semifree-ranging Barbary macaques (*Macaca sylvanus* L. 1758). *Folio primatol.*, **43**, 69–83.

Kummer, H. (1982) Social knowledge in free-ranging primates. In: *Animal Mind, Human Mind* (ed. D.R. Griffin), Springer, Berlin, pp. 113–32.

MacRoberts, M.H. and MacRoberts, B.R. (1966) The annual reproductive cycle of the

References

Barbary ape (*Macaca sylvana*) in Gibraltar. *Amer. J. Phys. Anthrop.*, **25**, 299–304.

Niemeyer, C.L. and Anderson, J.R. (1983) Primate harassment of matings. *Ethol. Sociobiol.*, **4**, 205–20.

Nigi, H., Tiba, T., Yamamoto, S., Floescheim, Y. and Ohsawa, N. (1980) Sexual maturation and seasonal changes in reproductive phenomena of male Japanese monkeys (*Macaca fuscata*) at Takasakiyama. *Primates*, **21**, 230–40.

Noordwijk, M.A. van (1985) Sexual behaviour of Sumatran long-tailed macaques (*Macaca fascicularis*). *Z. Tierpsychol.*, **70**, 277–96.

Paul, A. and Kuester, J. (1988) Life history patterns of Barbary macaques (*Macaca sylvanus*) at Affenberg Salem. In: *Ecology and Behaviour of Food-Enhanced Primate Groups* (eds. J.E. Fa and C.H. Southwick), Alan R. Liss Inc., New York, pp. 199–228

Roberts, M.S. (1978) The annual reproductive cycle of captive *Macaca sylvana. Folia primatol.*, **29**, 229–35.

Robinson, J.G. (1982) Intrasexual competition and mate choice in primates. *Amer. J. Primatol.*, **Suppl. 1**, 131–44.

Stephenson, G.R. (1975) Social structure of mating activity in Japanese monkeys. In: *Proc. Symp. 5th Congr. Int. Primatol. Soc.* (eds S. Kondo, M. Kawai, A. Ehara and S. Kawamura), Japan Science Press, Tokyo, pp. 63–115.

Simonds, P.E. (1973) Outcast males and social structure among bonnet macaques (*M. radiata*). *Amer. J. Phys. Anthrop.*, **38**, 599–604.

Takahata, Y. (1980) The reproductive biology of a free-ranging troop of Japanese monkeys. *Primates*, **21**, 303–29.

Takahata, Y. (1982) Social relations between adult males and females of Japanese monkeys in the Arashiyama B troop. *Primates*, **23**, 1–23.

Taub, D.M. (1980) Female choice and mating strategies among wild Barbary macaques (*Macaca sylvanus* L.). In: *The Macaques: Studies in Ecology, Behavior and Evolution* (ed. D.G. Lindburg), Van Nostrand Reinhold, New York, pp. 287–344.

Turkheim, G. de and Merz, E. (1984) Breeding Barbary macaques in outdoor open enclosures. In: *The Barbary Macaque: A Case Study in Conservation*, (ed. J.E. Fa), Plenum Press, New York, pp. 241–61.

CHAPTER SEVEN

Infant mistreatment in langur monkeys – sociobiology tackled from the wrong end?

Volker Sommer

7.1 INTRODUCTION

The transfer of infants between troop members, combined with the existence of a flamboyant natal coat, are characteristics of colobine monkeys (Horwich and Manski, 1975). Allomothering has been particularly well studied in both wild and captive Hanuman langurs (*Presbytis entellus*). Infants, strangely enough, are quite often mistreated by their caretakers; for example, they are made the object of a tug-of-war during transfers, are awkwardly carried, pressed on rocks, dragged, pushed out of trees or abandoned despite desperate screams (Jay, 1962; Mohnot, 1974; McKenna, 1975, 1981; Hrdy, 1977; Vogel, 1979, 1984; Scollay and De Bold, 1980; Dolhinow and Murphy, 1982).

Several hypotheses interpret infant transfer as an adaptive strategy which e.g. (a) frees the mother for foraging, (b) provides the infant with a potential for adoption, (c) provides agonistic buffering, or (d) the necessary training in infant care to the allomothers (cf. review in Hrdy, 1976). Some authors consider infant transfer to be (e) a fortuitous outcome of a general adaptation of infant care not selected for *per se* (Quiatt, 1979; Scollay and De Bold, 1980).

Assuming that infants suffer nutritional stress if handled excessively by nonmothers and considering the incidences of mistreatments, Wasser and Barash (1981) developed the hypothesis that (f) infant handling benefits the handlers by harming the handled infant since this reduces the number of resource competitors for their own offspring.

As a preliminary to a detailed revision of the functional hypotheses concerning infant transfer in langurs (Sommer and Vogel, in preparation) this

110

chapter tests the last (f) competition hypothesis against data collected during a long-term study of the Hanuman langurs of Jodhpur, India.

7.2 MATERIALS AND METHODS

A bisexual one-male troop of Hanuman langurs (*Presbytis entellus entellus* Dufresne, 1797) was studied from 25 October, 1981, until 20 December, 1982. The troop, denominated as Kailana-I (KI), lives near the city of Jodhpur, Rajasthan, India, in a partly protected open scrub habitat dominated by xerophytic plants such as *Prosopis juliflora, Acacia senegal* and *Euphorbia caducifolia.* The monkeys are predominantly terrestrial and habituated to humans due to provisioning by local people. There are no predators except feral dogs. For details on ecology and troop development see Winkler *et al.* (1984) and Sommer (1987).

All members of the troop were known individually. Between March and October 1982, six infants ($\male$7.5, $\male$1.4, $\female$4.5, $\male$11.3, $\female$12.1, $\male$6.4) were born. The infants are denominated first by the number of the mother, the second number indicating the birth rank since the study began in 1977. Only 11.3 and 12.1 were in fact the third and first infants of their mothers respectively, because $\female$11 and $\female$12 matured during the study. All infants died prematurely, most of them during infanticidal adult male replacements (Sommer, 1987).

The number of potential caretakers included:

1 adult male (the respective harem resident $\male$20, $\male$43, or $\male$11);

11 adult females (young $\female$11, $\female$12, $\female$13; middle-aged $\female$1, $\female$2, $\female$4, $\female$7, $\female$9; old $\female$3, $\female$6, $\female$8);

5 juvenile females ($\female$1.3, $\female$2.3, $\female$3.2, $\female$4.4, $\female$6.3);

3 juvenile males ($\male$7.4, $\male$8.2, $\male$11.2).

7.3 RESULTS

During their first six weeks of life, the infants spent 59.7% of the daytime with their mothers, 37.2% with non-mothers, and 3.0% alone (Table 7.1). Of the nonmothers, adult cycling females were responsible for 19.4% of all infant handling episodes (expectation 35.9%), pregnant and newly lactating females for 10.3% (expectation 21.3%), juvenile females for 69.7% (expectation 29.1%), and juvenile males for 0.4% (expectation 11.7%). The respective adult males never handled infants. In total, almost all infants were handled by all adult and nonadult females. Adult females occupying ranks 1–5 and 10–11 of the displacements hierarchy handled infants on average slightly less often than females with ranks 6–9; more striking, however, was the great variation from infant to infant and handler to handler. Caretakers mostly behaved warily. Occasionally, however, both mothers and nonmothers exhibited various forms of mistreatment including forcibly pulling off infants from

111

Table 7.1

Infant transfer in troop KI

Mother denotation	Rank[a]	Age	Infant	Period of observation		Daytime (%) spent			Allomaternal attempts	
				Days of life	Focal animal sample (h)	With mother	With allomother	Alone	Total[b]	% refused by mother
♀7	4	Middle-aged	♂7.5	1–7[c]	27.1	12.3	87.7	0.0	5	(40.0%)
♀1	7	Middle-aged	♂1.4	1–40[d]	94.2	51.5	44.3	4.2	111	(9.9%)
♀4	6	Middle-aged	♀4.5	1–29[e]	56.4	65.6	33.4	1.0	63	(3.2%)
♀11	1	Young	♂11.3	33–35[f]	10.9	64.5	22.8	12.7	12	(25.0%)
♀12	2	Primipar	♀12.1	2–33[g]	93.9	78.4	17.8	3.7	119	(27.7%)
♀6	10	Old	♂6.4	9–16[h]	36.0	58.2	39.5	2.3	86	(10.5%)
					Σ 318.5h	$\bar{x}$59.7	$\bar{x}$37.2	$\bar{x}$3.0	Σ396	$\bar{x}$(15.2%)

[a] Position in the displacement hierarchy at the time of birth.

[b] Both successful and abortive trials of taking an infant from its mother.

[c] Died on 10th day of life, probably due to starvation since its elder brother resumed suckling at mother's breast.

[d] Disappeared on 115th day of life, probably killed by a male.

[e] Disappeared on 278th day of life, probably killed by a male.

[f] Observer absent until 32nd day; male infanticide on 36th day.

[g] Disappeared on 98th day of life, probably killed by a male.

[h] Male infanticide on 47th day of life.

112

others, carrying in awkward positions other than ventral–ventral, holding upside down, mock-biting, pushing off, shoving down to the stomach, catapulting with feet – thus causing several falls out of trees and from buildings – or transporting the infants towards potential dangers such as dogs, humans or (partially infanticidal) male residents. Mostly, infants expressed their discomfort by screaming. However, they did not suffer any visible injuries.

Some typical mishandlings are illustrated by the following extracts from the field notes (see also Figure 7.1 and Figure 7.2):

(a) 16 April 1982, 17:13. Multipara ♀9 inspects 5-day-old ♀4.5 after taking the vocalizing baby from adult ♀13. ♀9 shoves the screaming infant to the ground, grimaces, and presses it with both feet against a stone. Juvenile ♀4.4 rescues her sibling. At the same time, juvenile ♀6.3 moves away from 19-day-old ♂1.4 who is taken by its mother. The baby tries to restore nipple contact but ♀1 resists, bites and forcibly pushes her screaming offspring into a muddy hole. The previous caretaker ♀6.3 returns and again takes the infant while its mother drinks at the nearby lake.

(b) 17 April 1982, 09:48. Together with her infant, ♀1 climbs a tree, displaces ♀2 and starts feeding. After three minutes, the screaming infant is bitten by its mother and pushed away. Before falling it is taken by juvenile ♀6.3 while the mother continues feeding.

(c) 24 April 1982, 13:06. ♀1 climbs down a tree where she dozed for about one hour with her infant. Suddenly, she pulls it off forcibly and solicits resident ♂20 for copulation. The infant is taken by its sister ♀1.3 while the mother consorts with the male.

(d) 27 April 1982, 09:04. ♀4 forcibly pulls off 16-day-old ♀4.5 and presses the screaming infant with one foot against the ground. The mother is obviously waiting until her daughter ♀4.4 takes the sibling. Subsequently, ♀4 feeds six minutes on the ground and afterwards at least 14 minutes up in a tree.

(e) 8 May 1982, 17:55. ♀4 causes a one metre fall of her baby ♀4.5 after pushing it off a wall. The mother feeds on a tree as soon as ♀2 takes the screaming infant. Four minutes later, ♀2 also throws it down one metre. The abandoned baby screams loudly for two minutes until the mother interrupts her feeding. The baby stops screaming as soon as she takes it.

(f) 12 August 1982, 11:28. ♀2 grooms ♀12 shortly before taking her 13-day-old infant ♀12.1 which is then transferred to juvenile ♀2.3. ♀12 and ♀2 mutually groom for nine minutes before the mother notices that juvenile

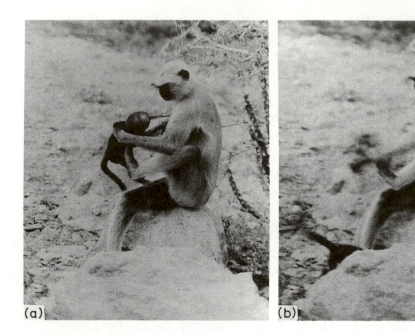

Figure 7.1 (a)(b) Despite its resistance. 9-day-old ♀12.1 is catapulted to the ground by multipara ♀4 who took the infant only two minutes before. (c) Primipara ♀12 retrieves her screaming offspring quickly but does not exhibit any aggression against the brutal caretaker (8 August 1982, 07:26).

♀6.3 transports her infant into a tree near the new infanticidal resident ♂43. Immediately, ♀12 pursues the caretaker, quickly retrieves her baby and moves away.

(g) 19 October 1982, 09:44. While carrying her 8-day-old brother ♂6.4, juvenile ♀6.3 approaches invader ♂46 within one metre. Immediately, mother ♀6 follows and chases the male away fiercely.

(h) 19 October 1982, 16:28. ♀6 rejects an attempt of ♀1 to take her infant, whereupon ♀ snatches ♀12.1 who crawls away after some seconds. After five minutes, ♀1 manages to take ♂6.4 without maternal resistance. Immediately, she shoves the baby from her belly to the ground. Ten seconds later, she retrieves the screaming baby, presses it with one foot on a rock and kicks it away again. ♀6, who fed a bit apart, retrieves her offspring after 20 seconds.

Of a total of 1981 handling episodes, infants were mistreated in 10.5% of all cases, i.e. by mothers in 70 out of 600 episodes (11.7%), by adult allomothers in 81 out of 485 episodes (16.7%), and by juveniles in 57 out of 896 episodes (6.4%). Tug-of-war transfers and clumsy carriage might reflect abilities lacking in the caretakers. Possibly harmful intentions are rather revealed through brute efforts to get rid of an infant and/or eventually abandoning it. The latter motivation was attributed to only 4.2% of all episodes. These are taken as the basis for further analysis (Table 7.2). Interindividual variability is considerable: 90.3% of all maternal and 52.8% of all nonmaternal mistreatments were suffered by infants ♂1.4 and ♀4.5. Although all nonmaternal caretakers mistreated infants at least once, only three individuals (♀1, ♀4, and ♀2, who had no infant of her own) were responsible for 54.7% of all incidences. In contrast to adult females, juveniles never got rid of infants brutally. Their mistreatments were restricted to mild rebuffing, letting them loose and moving away, or approaching potential dangers.

Although mothers rejected 15.2% of all allomaternal attempts (Table 7.1) it was not obvious that they did so to prevent subsequent mishandling. Firstly, in one third of all cases caretakers obtained infants prior to mistreatment from the mother itself (Table 7.3). Occasionally, rejected would-be caretakers started to groom the mother who consequently relaxed and permitted the transfer (see also field note (f)):

(i) 8 August 1982, 07:14. ♀1 quickly approaches ♀12 and tries to take her 9-day-old infant ♀12.1. The mother forcibly resist. ♀1 then briefly grooms ♀12 and can take the infant without further resistance.

Secondly, mothers did not discriminate against notorious mistreaters. On the contrary, the rather solicitous juveniles were rejected at a frequency above

115

Figure 7.2

Figure 7.2 (a) Though a new and possibly dangerous adult male sits close, 11-day-old ♂6.4 is abandoned (b), (c) five minutes after multipara ♀2 took him from his old mother ♀6.(d) After ten seconds, the screaming infant is rescued by juvenile ♀1.3 (22 October 1982, 09:26).

Table 7.2

Frequency of mistreatments during infant handling: pushing and pulling of, biting, shoving down, catapulting with feet, causing falls, abandoning

Infant	Mothers			Adult allomothers			Juvenile allomothers			Total		
	Handling episodes	Mistreatments (N)	(%)	Handling episodes	Mistreatments (N)	(%)	Handling episodes	Mistreatments (N)	(%)	Handling episodes	Mistreatments (N)	(%)
♂7.5	6		(0.0)	52	11	(21.2)	54	1	(1.9)	112	12	(10.7)
♂1.4	185	17	(9.2)	146	11	(7.5)	478	8	(1.0)	809	36	(4.4)
♀4.5	98	11	(11.2)	60	7	(11.7)	107	2	(1.9)	265	20	(7.5)
♂11.3	18		(0.0)	23	1	(4.3)	21		(0.0)	62	1	(1.6)
♀12.1	201	1	(0.5)	75	4	(5.3)	181		(0.0)	457	5	(1.1)
♂6.4	92	2	(2.2)	129	8	(6.2)	55		(0.0)	276	10	(3.6)
	Σ600	Σ31	x̄(5.2)	Σ485	Σ42	x̄(8.7)	Σ896	Σ11	x̄(1.2)	Σ1981	Σ84	x̄(4.2)

118

Table 7.3

From whom do caretakers obtain infants prior to mistreatments?

Source	Number of cases	
From mothers	17	(33.3%)
From half-sister	1	(2.0%)
From other allomothers	31	(60.8%)
Picking up unattended babies	2	(3.9%)
Total	51	

average. Maternal punishment was never observed (cf. Figure 7.1). Long-term sanctions against mistreaters could also not be detected. For example, ♀7 allowed ♀2 to handle and mishandle her infant ♂7.5 twenty-three times, although ♀2 had already abused her previous infant ♂7.3 two years before-hand (Vogel, personal communication). Mothers themselves rescued their babies in only 37.9% of all incidences (Table 7.4). Frequently, they ignored their screams or did not intervene in mishandlings although they were close by (field notes (e), (h)). There was no obvious correlation between the success rate of attempted retrievals and maternal rank. Two mothers at the opposite ends of their reproductive careers displayed above average concern: primipara ♀12 (64.7% of all rescues; field note (f)), and old ♀6 (75%; field note (g)) whose infant ♂6.4 was her last before reaching menopause. Further kin-oriented support was performed by half-sisters who carried out 8.3% of all rescues (field notes (a), (c), (d)).

For juvenile allomothers, the mean duration of handling prior to mistreatment was double that for the mean total duration. For adult allomothers, the duration was only slightly longer, and for mothers considerably shorter (Table 7.5).

The distribution of the abusive caretaker's activities just after getting rid of their charges was not random. All classes of caretakers subsequently engaged in allogrooming at a level above average. Sexual solicitation was also relatively frequent (field note (c)). Allomothers often went on handling another infant. Mothers showed a strong tendency to feed – especially on trees (field notes a, b, d, e) – whereas monitoring (an activity not combined with manipulation or locomotion) – was greatly underrepresented (Table 7.6).

7.4 DISCUSSION

Infant transfer in langurs is a very complex phenomenon. Any attempt to point out behavioural trends can easily be matched by an example illustrating

119

Table 7.4
Who rescues mistreated infants ?[a]

Rescuer	Number of cases	
Mothers	55	(37.9%)[b]
Half-sisters	12	(8.3%)
Adult allomothers	24	(16.6%)
Juvenile allomothers	54	(37.2%)
Total	145	

[a] Rescue implies taking vocalising or abandoned infants from the caretaker or the ground, or intercepting the passage of a caretaker towards a potential danger.
[b] Range 0.0–75.0% for mothers of six different infants.

Table 7.5
Duration of infant handling (episodes with known duration only)

Caretaker	Total handlings		Prior mistreatment	
	Episodes	Mean duration (min)	Episodes	Mean duration ($\pm$ SD) (min)
Mothers	597	18.7	25	12.7 $\pm$ 17.7
Adult allomothers	475	3.5	41	4.7 $\pm$ 5.5
Juvenile allomothers	885	4.3	9	8.1 $\pm$ 4.9

the opposite due to considerable variability both within age-sex-classes of caregivers as well as of given individuals over a succession of infants (Dolhinow and Krusko, 1984).

The recent findings confirm Hrdy's dichotomous slogan of 'brutal multi-paras; solicitous, wary nulliparas' (1977) because juveniles not only handled infants above expectation frequencies but treated them more carefully. At Jodhpur, the handling experience resulted in mistreatment of the infants in every 81st episode with a juvenile caretaker compared with every 11th episode with an adult allomother. More surprisingly, mothers mistreated their own infants in every 19th episode. Previously, only Hrdy (1977) and Scollay and De Bold (1980) mentioned occasional clumsiness of mothers.

The competition hypothesis that infant handling benefits the handler by harming the handled infant and in this way reduces resource competition for own offspring leads to several predictions (Wasser and Barash, 1981; Wasser, 1983) which can be tested against the current data:

Discussion

Table 7.6

Activity of caretakers subsequent to infant mistreatment

Activity	Expected percentage[a]	Observed percentage subsequent mistreatments	
		Allomaternal episodes (N=27)	Maternal episodes (N=25)
Dozing	24.7	0.0	0.0
Monitoring	23.7	25.9	4.0
Foraging	23.5	22.2	68.0
(on the ground)		(3.7)	(24.0)
(on trees)		(11.1)	(32.0)
(drinking)		(0.0)	(8.0)
(provisioning)		(7.4)	(4.0)
Locomotion	9.3	0.0	0.0
Allogrooming	7.6	11.1	12.0
Infant handling[b]	—	37.0	4.0
Solicitation for copulation	<1	3.7	12.0
Others	10.0	0.0	0.0

[a] Monthly mean during a one-year study of the neighbouring troop B living under similar ecological conditions (Winkler, 1981: 99, 117).

[b] Both own and alien infants; this category was not used by Winkler, (1981).

1. Newborns (0–3 months old) should be handled more than older infants because the former are more vulnerable to the nutritional consequences of maternal separation, as well as less able to defend themselves and/or run away.

Confirmation of prediction: inconclusive. Certainly, the frequency of hand-ling decreases with growing age and physical independence. However, as an unavoidable correlate of ontogeny, this is probably true for *all* primates having infant transfer systems (e.g. Dolhinow and Murphy, 1982 for *P.entellus*; Sommer, 1984 for *Callithrix jacchus*). Consequently, it is impossible to falsify the hypothesis, and its confirmation has no intrinsic value.

Maternal separation is life-threatening for langur neonates when exceeding a certain limit. Infant ♂7.5 spent 87.7% of the daytime during its first week of life away from its mother and ultimately died, probably due to starvation. However, this was not a result of extensive allomothering but of a lack of retrievals by its mother: the milk meant for the neonate was drunk by the elder brother ♂7.4. ♀7 obviously did not perceive a need to suckle her newborn. Other infants spent up to 48.5% of the daytime apart from their mothers and developed normally. As demonstrated by all quantitative studies to date, extended separation is a normal pattern in langurs and does not jeopardize the infant. Perhaps, colobine mothers compensate for infant transfer by producing richer milk (Hrdy, 1977).

2. Both nulliparous and multiparous females should handle infants, and treat them roughly.

Confirmation of prediction: tendency positive. Since both juveniles and adult females handled and mistreated infants, not lack of experience but intention seems to be the key factor for abusive behaviour.

3. Females of some reproductive states – specifically, pregnant and newly lactating individuals – should handle infants more than others because their (prospective) infants would have the most gain from their behaviour.

Confirmation of prediction: tendency negative. Adult fertile females abused infants relatively more frequently than did juveniles. However, with regard to the total frequency of handling, juveniles took infants twice as frequently than expected. Moreover, in contradiction to the findings of Hrdy (1977) and Vogel (1984), pregnant females and lactating mothers were half as frequently involved as expected (Sommer, 1985).

4. Mothers should resist the efforts of others attempting to handle infants, especially those who would receive the most benefit from mistreating them.

Confirmation of prediction: tendency negative. Although mothers rejected 15.2% of all allomaternal attempts, they did not discriminate against well known mistreaters despite the fact that two of them (♀1, ♀4) were lactating individuals which, according to the prediction, would receive the highest gain. Moreover, even notorious mishandlers received subsequently abused infants in one third of all cases from the mothers themselves. Despite previous rejection, a mother would allow transfer if the attempted caretaker started grooming her (field note (i); Jay, 1963 McKenna 1975, for *P. entellus*; Rowell *et al.*, 1964, for *Macaca mulatta*; Struhsaker, 1971, for *Cercopithecus aethiops*).

5. High-ranking females should handle infants more than low-ranking females do.

Confirmation of prediction: negative. Juveniles who are not yet integrated into the hierarchy, handle most. Adult females of ranks 1–5 are clearly less engaged than females of ranks 6–9. The most brutal caretakers (♀1, ♀4, ♀2) held 4th–8th positions in the rank order.

6. Infants of low-ranking mothers should be handled more frequently than infants of high-ranking mothers.

Confirmation of prediction: indecisive. The prediction seems to draw some support since the three highest ranking mothers refused more allomaternal attempts than lower ranking ones, and the infants of the two highest ranking mothers suffered the fewest mistreatments. However, a strict correlation between maternal rank and the time spent with allomothers was not detectable. Similarly, in agreement with Scollay and De Bold (1980) and Dolhinow and Murphy (1982), there was no correlation between dominance position and rate of successful retrievals.

Vogel (1984) emphasized that the competition hypothesis calls for an additional prediction:

7. To avoid diminution of a handler's reproductive fitness, closely related infants should not be handled or mistreated.

Confirmation of prediction: negative. Caretakers preferred to handle half- or fullsiblings (Sommer, unpublished; Vogel, 1979, 1984; Cf. differing findings of Hrdy, 1977; Dolhinow and Krusko, 1984). More important, mothers themselves did not hesitate to mishandle their own infants, thus jeopardizing their personal fitness *if* the observed mistreatments were in fact likely to harm their offspring.

In this regard, it is important to note that reports of actual injuries caused by allomothering were restricted to small cuts and scrapes of the head, nose and tail at the most (Mohnot, 1974; Hrdy, 1977; McKenna, 1981; Dolhinow and Murphy, 1982; Vogel, 1984). Langur babies are obviously extraordinary robust. Consequently, their mothers might maintain other standards for adequate treatment as do human observers, both in terms of tolerating separation as in evaluating the dangers of rough handling. Most reports except Jay (1962) agree that mothers frequently ignore their baby's vocalizations and do not retrieve them (Sugiyama, 1965; Mohnot, 1974; Hrdy, 1977: Scollay and De Bold, 1980). The double standard becomes apparent if a 'real' danger (such as a human, unfamiliar adult male, or dog) draws near since in such instances mothers tend to retrieve their infant very quickly (field notes (f), (g); Sugiyama, 1965; Hrdy, 1977).

Vogel's statement of 'at least two infant-losses (deaths) in consequence of "allomothering"' (1984, see also 1979) could not stand up against close scrutiny of the incidents since the fault of nonmothers was not definitely provable (personal communication). Although the possibility of infant deaths in connection with mishandling by allomothers should not be entirely excluded, such fatalities might likewise be caused by the mothers themselves. (Mohnot, 1977, assumes that inexperienced mothers or nonmothers choked infants to death.) This aspect, however, would turn the discussion in a completely reverse direction.

The competition hypothesis is hence rejected for langurs in agreement with Vogel (1984) This is not to deny its importance for other species since Wasser (1983) was able to confirm its predictions for *Papio cynocephalus* and Silk (1980) for *Macaca radiata*. However, basic differences exist between the social set-ups of langurs and baboons or macaques. The latter cercopithecids are organized in multiple matrilines with comparably low degrees of relatedness between individuals of different matriarchies. The thresholds of female–female aggression are low and reflect intense inter-lineage competition. Langur troops at Jodhpur with their one-male breeding structure, on the other hand, consist of closely related lineages because females of the same age-cohorts are paternal half-sisters (Sommer, unpublished; Hrdy and Hrdy, 1976). Consequently, female hierarchies are flexible and age-dependent and the troop members are expected to exhibit close co-operation and kin-support (see also Hrdy, 1981). A special case might be 'aunting-to-death' due to infant kidnapping *between* langur troops (Mohnot, 1980), perhaps an expression of territorial resource competition.

Lancaster (1971), for *Cercopithecus aethiops*) and Quiatt (1979), for *Macaca mulatta*) assume that mothers punish clumsy allomothers who will thus learn proper infant treatment. For langurs, we can exclude such a process of mutual conditioning due to the lack of maternal sanctions against mishandlers (Mohnot. 1974; McKenna, 1981).

Since abusive infant handling is widespread throughout the langur study sites, it can hardly be interpreted as pathologically similar to the behaviour of 'motherless' rhesus monkeys (Harlow, 1971). After all, what might be its functional background?

The prolonged duration of handling by allomothers prior to mistreatment and the tendency to nevertheless handle another infant just after getting rid of the previous charge suggests that allomothers were simply bored with a particular infant. It is only because infants cling fiercely to their carriers, that the latter's efforts assume violent traits: 'It is this conflict of interest between an allo-mother who wishes to shed an encumbrance and an infant that clings which underlines the brutality . . .' (Hrdy, 1977).

Additionally, a caretaker may make an infant cry in order to attract another troop member who takes over the burden (field note (d)). Similarly, abandoning, retrieving and renewed abandoning of a vocalizing infant might reflect the conflict of a rough caretaker who feels uncomfortable if a scream-ing infant is close by (field note (h)). At Galta near Jaipur (Rajasthan) care-takers developed an alternative tactic to reinstall peace: As soon as they sit on their charges, the infants stop screaming (A. Lobo, personal communication). Obviously, langurs can elicite a large number of speculations in puzzled primatologists.

The fact that caretakers at Jodhpur after mistreatments engaged more often than expected with other troop members in allogrooming might indicate their need for some reassurance or appeasement after having had an unpleasant inter-action with an infant. Similarly, the urge for sexual solicitation as a suddenly arising hormonal stimulus may sometimes entail enforced efforts to abandon an infant. On the part of mothers, feeding seems to be the primary reason to dispose of their offspring. This confirms the hypothesis that infant transfer is an adaptation which ensures the mother's nutritional input by freeing her for foraging (field notes (a), (b), (d), (e), (h)) and enables her to maintain social contacts (field notes (c), (f)).

SUMMARY

Assuming that langur infants suffer nutritional stress due to the common pattern of excessive infant transfer and with regard to the incidences of mistreatments, the competition hypothesis states that infant handling benefits the handler by harming the handled infant since this reduces the number of resource competitors for own offspring. However, amongst free-ranging langurs at Jodhpur, India, nonmothers as well as mothers themselves engaged in abusive behaviours. Neither the transfer itself nor rough treatment jeopardized the infant. Brutal handling resulted mostly from efforts to get rid of the infant when caretakers attempted to free themselves for foraging or social contacts.

ACKNOWLEDGEMENTS

The field study was funded by the Deutscher Akademischer Austauschdienst (DAAD) and the Government of India, Department of Education and Social Welfare, under its Indo–German Cultural Exchange Programme. Presently, financial assistance is provided by the Feodor Lynen Programme of the Alexander von Humboldt Foundation. I am thankful to S.M. Mohnot and C. Vogel for their constant support.

REFERENCES

Dolhinow, P. and Krusko, N. (1984) Langur monkey females and infants: The female's point of view. In: *Female Primates: Studies by Women Primatologists,* (ed. M.S. Small), Alan R. Liss, New York, pp. 37–58.

Dolhinow, P. and Murphy, G. (1982) Langur monkeys (*Presbytis entellus*) development: the first 3 months of life. *Folia primatol.,* **39**, 305–31.

Harlow, H.F. (1971) *Learning to Love,* Ballantine, New York.

Horwich, R.H. and Manski, D. (1975) Maternal care and infant transfer in two species of colobus monkeys. *Primates,* **16**, 49–73.

Hrdy, S.B. (1976) Care and exploitation of nonhuman primate infants by conspecifics other than the mother. In: *Advances in the Study of Behavior, Vol. 6* (eds J.S. Rosenblatt, R.A. Hinde, E.A. Shaw, and C. Beer), Academic Press, New York, pp. 101–56.

Hrdy, S.B. (1977) *The Langurs of Abu,* Harvard University Press, Cambridge, MA.

Hrdy, S.B. (1981) "Nepotists" and "altruists": The behavior of old females among macaques and langur monkeys. In: *Other Ways of Growing Old* (eds P.T. Amors and

S. Harrell) Stanford University Press, Stanford, pp. 59–76.

Hrdy, S.B. and Hrdy, D.B. (1976) Hierarchical relations among female Hanuman langurs (Primates: Colobinae, *Presbytis entellus*) *Science*, **193**, 913–15.

Jay, P.C. (1962) Aspects of maternal behavior among langurs. *Ann. N. Y. Acad. Sci.*, **102**, 468–76.

Jay, P.C. (1963) The Social Behavior of the Langur Monkeys, Thesis, University of Chicago.

Lancaster, J. (1971) Play-mothering: the relations between juvenile females and young infants among free-ranging vervet monkeys (*Cercopithecus aethiops*). *Folia primatol.*, **15**, 161–82.

McKenna, J.J. (1975) *An Analysis of the Social Roles and Behavior of Seventeen Captive Hanuman Langurs (Presbytis entellus)*. Thesis, University of Oregon.

McKenna, J.J. (1981) Primate infant caregiving behavior. In: *Parental Care in Mammals* (eds D.J. Gubernick and P.H. Klopfer), Plenum Press, New York, pp. 389–416.

Mohnot, S.M. (1974) Ecology and Behavior of the Common Indian Langur, Presbytis entellus, in India. Thesis, University of Jodhpur.

Mohnot, S.M. (1977) Observations on maternal behavior in the langur, *Presbytis entellus*, in India. In: *The Natural Resources of Rajasthan* (ed. M.L. Roonwal), Jodhpur University Press, Jodhpur, pp. 505–14.

Mohnot, S.M. (1980) Intergroup infant kidnapping in Hanuman langur. *Folia primatol.*, **34**, 259–77.

Quiatt, D. (1979) Aunts and Mothers: Adaptive implications of allomaternal behavior of nonhuman primates. *am. Anthropol.*, **81**, 310–19.

Rowell, T.E., Hinde, R.A. and Spencer-Booth, Y. (1964) "Aunt"–infant interaction in captive rhesus monkeys. *J. Anim. Behav.*, **12**, 219–26.

Scollay, P.A. and De Bold, P. (1980) Allomothering in a captive colony of Hanuman langurs (*Presbytis entellus*). *Ethol. Sociobiol.*, **1**, 291–9.

Silk, J.B. (1980) Kidnapping and female competition among captive bonnet macaques. *Primates*, **21**, 100–110.

Sommer, V. (1984) Dynamics of group structure in a family of the common marmoset, *Callithrix jacchus* (Callitrichidae). In: *Current Primate Researches* (eds M.L. Roonwal, S.M. Mohnot and N.S. Rathore), Jodhpur University Press, Jodhpur, pp. 315–33.

Sommer, V. (1985) Weibliche und männliche Reproduktionsstrategien der Hanuman-Languren (*Presbytis. entellus*) von Jodphur (Rajasthan/Indien), Thesis, University of Göttingen.

Sommer, V. (1987) Infanticide among free-ranging langurs (*Presbytis entellus*) at Jodhpur (Rajasthan/India): Recent observations and a reconsideration of hypotheses. *Primates*, **28**, 163–97.

Struhsaker, T.T. (1971) Social behavior of mother and infant vervet monkeys (*Cercopithecus aethiops*). *Anim. Behav.*, **19**, 233–50.

Sugiyama, Y. (1965) Behavioral development and social structure in two troops of Hanuman langurs (*Presbytis entellus*). *Primates*, **6**, 213–47.

Vogel, C. (1979) Der Hanuman Langur (*Presbytis entellus*), ein Paradeexempel für die theoretischen Konzepte der Soziobiologie? *Verh. Dtsch. Zool. Ges.*, **1979**, 73–89.

Vogel, C. (1984) Pattern of infant transfer within two troops of common langurs (*Presbytis entellus*) near Jodhpur: Testing hypotheses concerning the benefits and risks. In:

Reference

Current Primate Researches, (eds M.L. Roonwal, S.M. Mohnot and N.S. Rathore), Jodhpur University Press, Jodhpur, pp. 361–80.

Wasser, S.K. (1983) Reproductive competition and cooperation among female yellow baboons. In: *Social Behavior of Female Vertebrates* (ed. S.K. Wasser), Academic Press, New York, pp. 349–90.

Wasser, S.K. and Barash, P.P. (1981) The selfish "allomother": a comment on Scollay and De Bold. *Ethol. Sociobiol.*, **2**, 91–3.

Winkler, P. (1981) Zur öko-Ethologie freilebender Hanuman-Languren (*Presbytis entellus entellus* Dufresne, 1797) in Jodhpur (Rajasthan), Indien. Thesis, University of Göttingen.

Winkler, P., Loch, H. and Vogel, C. (1984) Life history of Hanuman langurs (*Presbytis entellus*): reproductive parameters, infant mortality, and troop development. *Folia primatol.*, **43**, 1–23.

PART TWO

Sociobiological aspects of human sexuality and reproductive strategies

CHAPTER EIGHT

Human male–female differences in sexual desire

Donald Symons and Bruce Ellis

8.1 INTRODUCTION

All psychological theories, even the most extreme empiricist/environmental-ist/associationist ones, imply a human nature; that is to say, they imply that some brain/mind mechanisms are typical of *Homo sapiens* as a species, in the sense that having arms rather than wings is typical of *Homo sapiens*. Theories differ, however, in the extent to which these species-typical brain/mind mechanisms are conceived of as few and generalized, on the one hand, or many and specialized, on the other. Darwinism strongly favours the latter. Organisms are designed by natural selection to solve specific problems; there is no more reason to imagine that one or a few generalized brain/mind mechanisms could solve all behavioural problems than there is to imagine that one or a few generalized organs could solve all physiological problems (Symons, 1987b). A corollary of the basic Darwinian expectation that the human brain/mind comprises specialized mechanisms is the expectation that the human brain/mind is sexually dimorphic: the nature of mammalian reproduction ensures that throughout the course of evolutionary history hominid males and females encountered very different reproductive oppor-tunities and constraints, hence selection can be expected to have designed males and females to solve somewhat different problems. In other words, evolution-mindedness leads us to expect intraspecific male–female brain differences for precisely the same reason that it leads us to expect interspecific brain differences.

Evolution-mindedness can even be a useful guide to forming specific hypotheses about the natures of sexually dimorphic psychological mechan-isms. Whatever typical parental investments (Trivers, 1972) may have been during the course of human evolutionary history, ancestral males and females must have differed enormously in the minimum possible investment. Ancestral males potentially could have benefited reproductively from copulating with any fertile female as long as the risks were low enough, hence it is reasonable to hypothesize that selection favoured males who found new females – in Byron's words, 'fresh features' – especially sexually attractive. Ancestral females, on the other hand, would have had nothing to gain reproductively and a great deal to lose from random copulations with new males, hence selection is unlikely to have favoured females who were sexually attracted to males on the basis of variety *per se* or merely because the males were there (Symons, 1979, 1987a).

In sum, evolution-mindedness leads us to expect that among human beings the experience of sexual attraction is underpinned by a number of specialized psychological mechanisms, some of which are sexually dimorphic; in particular, partner variety *per se* should be important to males but not to females. But how can such mechanisms be identified and described? Human action must typically result from the simultaneous interaction of many psychological mechanisms, most of which, presumably, are sexually monomorphic. Consider, for example, the problem of using data on frequency of sex with new partners to illuminate male–female differences in the psychology of sexual attraction. Each time a man has sexual intercourse with a new woman a woman is having sexual intercourse with a new man, hence the numerical logic of heterosexual coupling ensures that males and females cannot differ in the total number of new partners each sex has. One way around this problem is to examine the literature on the sex lives of homosexuals, which suggests that men and women differ dramatically in the significance of partner variety for sexual attraction. Until very recently, many – probably most – male homosexuals in the US had sexual relations primarily with strangers and thus had hundreds or thousands of sexual partners; lesbians, however, had sexual relations primarily with steady partners and averaged no more sexual partners than heterosexual women did (Symons, 1979, 1980). Because of the fear of AIDS, however, many homosexual men have drastically reduced the frequency with which they have sexual relations with new partners. Obviously, this is evidence, not that fresh features have suddenly become less attractive to homosexual men, or that males and females have suddenly become similar in their sexual psychologies, but rather that sexual activities are determined by many factors in addition to the psychology of sexual attraction. 'After all,' remarks Wilson (1987), 'manifest libido fell quite markedly in Cambodia when the Khmer Rouge regime applied the death penalty to premarital sex and flirting.'

Hypotheses about male–female differences in the psychology of sexual attraction will eventually be tested by the neurosciences. In the meantime, however, other kinds of evidence can be obtained. One approach is to examine the ethnographic record of sexuality; cross-cultural regularities may provide clues to male and female natures (Symons, 1979). A second approach, as mentioned above, is to study the sex lives of homosexuals; hetero-sexual activities, other than rape, reflect compromises of male and female dispositions, but the sex lives of homosexual men and women provide insight into male and female sexualities in their uncompromised states (Symons, 1979, 1980). Still another approach is to compare male and female literatures of erotic fantasy; that is, to compare visual pornography, which is consumed almost entirely by men, with the romance novel, which is consumed almost entirely by women. A fourth approach is to design questionnaires specifically to illuminate male–female differences.

Since sexual intercourse exposes heterosexual males and females to very different risks, the mere fact of sex differences in activities, or even in willing-ness to engage in activities, is not conclusive evidence for sex differences in the mechanisms of attraction. Men and women might act differently because of the operation of sexually monomorphic cognitive mechanisms of risk assessment; therefore, comparing male and female fantasies is more likely to reveal sex differences in desire than comparing male and female activities is (Wilson, 1987). Along these lines, Symons (1979), following Sigusch and Schmidt (1971), predicted the existence of a major sex difference in the responses of married people to the following question: 'If you had the oppor-tunity to copulate with an anonymous member of the opposite sex who was as physically attractive as your spouse but no more so, and as competent a lover as your spouse but no more so, and there was no risk of discovery, disease, or pregnancy, and no chance of forming a more durable liaison, and the copu-lation was a substitute for an act of martial intercourse, not an addition, would you do it?' The intent of this question was to eliminate, in the imagin-ation of male and female respondents alike, the primary real-life risks associated with sexual intercourse in the hope that responses to this question would provide more insight into the psychology of sexual attraction than data on sexual activities do. Since, in the intervening years, no one to our know-ledge has included this or a similar question in a questionnaire, we decided to do so. We also included questions to investigate the effects of varying the physical attractiveness of the imagined partner and the chance to form a more durable relationship with the imagined partner. We expected that physical attractiveness would be particularly important to men and that the chance to form a more durable relationship would be particularly important to women (Buss and Barnes, 1986; Symons, 1979, 1987a).

8.2 METHODS

The subjects were 415 students (232 females, 183 males) enrolled in intro-ductory-level general education courses (General Psychology, Introduction to Sociology, and Cultural Anthropology) at a highly-rated California state university (with a predominantly middle-class, white student body drawn from all parts of the state) and at an urban junior college (with a pre-dominantly local, largely working-class student body consisting primarily of racial and ethnic minorities). Approximately half of the subjects came from each school. Introductory-level general education courses were chosen for study because the students enrolled in such courses represent a broad cross section of academic majors at both schools. The subjects included 200 females and 152 males in the 17–24 age group, 15 females and 17 males in the 25–29 age group, 11 females and 9 males in the 30–39 age group, and 6 females and 1 male in the 40 and above age group (four males did not state their ages); 58.5% of the subjects were Caucasian, 20.1% were Asian, 7.8% were Hispanic, 7.3% were Black, and 6.3% were 'other'. The overwhelming majority (89.4%) had never been married. Data were collected in the fall of 1986.

The questionnaire was an anonymous paper-and-pencil survey consisting of 70 multiple-choice questions; in this article, however, we will consider only the first 12 questions, which constituted an independent section of the survey. Subjects who indicated that they currently had a steady partner (whether or not they were married) were instructed to answer questions 1(P) through 6(P); subjects without steady partners answered a very similar set of questions, 1(NP) through 6(NP). Responses were recorded on a scantron-like sheet designed to minimize the visibility of answers and ensure confidentiality.

The survey was given to entire class sections without prior warning in order to obtain as large a response rate as possible. To engage the subjects interest in the study, they were told, before the questionnaires were dis-tributed, that the experimenter would return to their class at a later date to explain the purpose of the survey and to discuss the results (which he did). Interest in the survey ran high; although participation was, of course, completely voluntary, virtually every class member chose to participate. Subjects placed their completed answer sheets in a large envelope at the front of the classroom. Five per cent of the surveys were discarded because of incompleteness, internal contradictions, or indications of carelessness.

8.3 RESULTS

The response to questionnaire items of people with and without steady part-ners are summarized in Tables 8.1 and 8.2 respectively. (Question 1(P) is an abbreviated version of the original question that Symons (1979) predicted would elicit a substantial sex difference.) On only one [5(P)] of twelve ques-tions did sex differences in responses fail to reach statistical significance at the 0.05 level, and in most cases sex differences were dramatic. For example, males with steady partners were four times as likely as females with steady partners to answer 'certainly would' to question 1(P), and females were two and a half times as likely as males to answer 'certainly not' to this question. Sex differences were even greater in responses to the corresponding question

[1(NP)] presented to people without steady partners: males were six times as likely as females to answer 'certainly would', while females were more than two and a half times as likely as males to answer 'certainly not'.

Since the questions presented to people with steady partners [1(P)–6(P)] were not identical to the corresponding questions presented to people without steady partners [1(NP)–6(NP)], we did not feel justified in formally comparing the responses given by these two groups. Nevertheless, a casual comparison of Table 8.1 with Table 8.2 strongly suggests that, regardless of sex, people with steady partners were less willing than people without steady partners to have sexual intercourse with a new person.

To analyze the importance of the physical attractiveness of the imagined partner, we compared the responses of individuals in pairs of questions that differed only with respect to the physical attractiveness of the imagined partner. Table 8.3 summarizes these comparisons. For example, the first line of Table 8.3 compares each individual's responses to question 1(P) with his or her response to question 2(P) for all individuals who answered both questions. (Questions 1(P) and 2(P) are identical except that the imagined partner is less physically attractive in the latter.) The entry in the first column (percentage changed) indicates that 56.3% of the males who responded to both questions changed their responses from question 1(P) to question 2(P). The entry in the second column (number more likely) indicates that no males changed in the direction of being more likely to have sexual intercourse with the imagined partner; the entry in the third column (number less likely) indicates that 49 males changed in the direction of being less likely to have sexual intercourse with the imagined partner. The fourth column (p) gives the results of an ordinary sign test using the procedure for paired samples (Gibbons, 1976). In this case, the null hypothesis – that response changes from $1(p)$ to $2(p)$ were as likely to be in one direction as the other – can be confidently rejected.

The data summarized in Table 8.3 suggest that reducing the imagined partner's physical attractiveness had a much greater effect than increasing the imagined partner's physical attractiveness did. Both males and females were more likely to change their responses if the physical attractiveness of the imagined partner was reduced than they were if the physical attractiveness of the imagined partner was increased. These data also suggest that the physical attractiveness of the imagined partner may be more important to males than females. Decreasing or increasing the physical attractiveness of the imagined partner caused a higher percentage of males than females to change their responses. The male–female difference was especially striking when the physical attractiveness of the imagined partner was increased: if they changed at all, males overwhelmingly changed in the intuitively expected direction; that is, when the physical attractiveness of the imagined partner was increased, males who changed in the 'more likely' direction significantly outnumbered males who changed in the 'less likely' direction. But in only one [1(NP)–

135

Table 8.1

Responses of people with steady partners to questionnaire items

---

1(P)* If the opportunity presented itself of having sexual intercourse with an anonymous member of the opposite sex who was as competent a lover as your partner but no more so, and who was as physically attractive as your partner but no more so, and there was *no* risk of pregnancy, discovery, or disease, and *no* chance of forming a more durable relationship, do you think you would so?

$[\chi^2(3) = 28.15, p = 0.0000]$

	Certainly would	Probably would	Probably not	Certainly not
Females				
No.	5	16	40	61
Percent	4.1	13.1	32.8	50.0
Males				
No.	15	25	30	17
Percent	17.2	28.7	34.5	19.5

2(P) What if the person was somewhat *less* physically attractive than your partner?

$[\chi^2(3) = 11.24, p = 0.0105]$

	Certainly would	Probably would	Probably not	Certainly not
Females				
No.	0	7	38	77
Percent	0	5.7	31.1	63.1
Males				
No.	1	14	34	38
Percent	1.1	16.1	39.1	43.7

3(P) What if the person was somewhat *more* physically attractive than your partner?

$[\chi^2(3) = 37.77, p = 0.0000]$

	Certainly would	Probably would	Probably not	Certainly not
Females				
No.	7	17	37	61
Percent	5.7	13.9	30.3	50.0
Males				
No.	20	30	23	14
Percent	23.0	34.5	26.4	16.1

4(P) Suppose there *was* a chance of forming a more durable relationship with this 'anonymous' person. And suppose that you overheard a friend describe this person as: A bright and generous individual, someone with a sense of humor who shares their time freely and is well-liked by others. He/she is a successful architect, is fond of children, likes to travel, and is a serious amateur runner. Now suppose that again the opportunity presented itself of having sexual intercourse with this person. Do you think, if this person was as physically attractive as your partner but no more so, that you would do so?

$[\chi^2(3) = 16.73, p = 0.0008]$

	Certainly would	Probably would	Probably not	Certainly not
Females				
No.	9	17	50	46
Percent	7.4	13.9	41.0	37.7
Males				
No.	11	28	33	15
Percent	12.6	32.2	37.9	17.2

5(P) What if this person was somewhat *less* physically attractive than your partner?
$[\chi^2(3) = 6.91, p = 0.0750]$

	Certainly would	Probably would	Probably not	Certainly not
Females				
No.	2	10	48	61
Percent	1.7	8.3	39.7	50.4
Males				
No.	4	14	38	31
Percent	4.6	16.1	43.7	35.6

6(P) What if this person was somewhat *more* physically attractive than your partner?
$[\chi^2(3) = 25.96, p = 0.0000]$

	Certainly would	Probably would	Probably not	Certainly not
Females				
No.	6	19	47	46
Percent	5.1	16.1	39.8	39.0
Males				
No.	20	23	29	13
Percent	23.5	27.1	34.1	15.3

* P stands for 'partner'; i.e., 1(P) is the first question presented to individuals with steady partners.

Table 8.2
Responses of people without steady partners to questionnaire items

1(NP)* If the opportunity presented itself to have sexual intercourse with an anonymous member of the opposite sex who was as physically attractive as yourself but no more so, and there was *no* chance of forming a more durable relationship, and *no* risk of pregnancy, discovery, or disease, do you think you would do so?
[$\chi^2(3) = 38.30, p = 0.0000$]

	Certainly would	Probably would	Probably not	Certainly not
Females				
No.	6	28	33	42
Percent	5.5	25.7	30.3	38.5
Males				
No.	32	30	24	10
Percent	33.3	31.3	25.0	10.4

2(NP) What if the person was somewhat *less* physically attractive than yourself?
[$\chi^2(3) = 34.32, p = 0.0000$]

	Certainly would	Probably would	Probably not	Certainly not
Females				
No.	0	9	45	56
Percent	0	8.2	40.9	50.9
Males				
No.	8	28	39	20
Percent	8.4	29.5	41.1	21.1

3(NP) What if the person was somewhat *more* physically attractive than yourself?
[$\chi^2(3) = 41.81, p = 0.0000$]

	Certainly would	Probably would	Probably not	Certainly not
Females				
No.	9	30	36	35
Percent	8.2	27.3	32.7	31.8
Males				
No.	38	33	16	8
Percent	40.0	34.7	16.8	8.4

4(NP) Suppose there *was* a chance of forming a more durable relationship with this 'anonymous' person. And suppose that you overheard a friend describe this person as: 'A bright and generous individual, someone with a sense of humor who shares their time freely and is well-liked by others. He/she is a successful architect, is fond of children, likes to travel, and is a serious amateur runner.' Now suppose that again the opportunity presented itself to have sexual intercourse with this person. Do you think, if the person was as physically attractive as yourself but no more so, that you would do so?
$[\chi^2(3) = 19.65, p = 0.0002]$

	Certainly would	Probably would	Probably not	Certainly not
Females				
No.	12	41	33	23
Percent	11.0	37.6	30.3	21.1
Males				
No.	30	41	14	10
Percent	31.6	43.2	14.7	10.5

5(NP) What if the person was somewhat *less* physically attractive than yourself?
$[\chi^2(3) = 11.35, p = 0.0100]$

	Certainly would	Probably would	Probably not	Certainly not
Females				
No.	3	27	47	33
Percent	2.7	24.5	42.7	30.0
Males				
No.	8	38	31	17
Percent	8.5	40.4	33.0	18.1

6(NP) What if the person was somewhat *more* physically attractive than yourself?
$[\chi^2(3) = 30.72, p = 0.0000]$

	Certainly would	Probably would	Probably not	Certainly not
Females				
No.	12	39	35	24
Percent	10.9	35.5	31.8	21.8
Males				
No.	39	34	12	10
Percent	41.1	35.8	12.6	10.5

*NP stands for 'no partner'; i.e., 1(NP) is the first question presented to individuals without steady partners.

Table 8.3

Effects of varying physical attractiveness of imagined partner on responses to questionnaire items

Response comparisons	Males				Females			
	% changed	Number more likely	Number less likely	p	% changed	Number more likely	Number less likely	p
Less attractive								
1(P)–2(P)	56.3	0	49	<0.05	25.4	1	30	<0.05
4(P)–5(P)	43.7	1	37	<0.05	24.6	0	29	<0.05
1(NP)–2(NP)	52.6	0	50	<0.05	42.2	3	43	<0.05
4(NP)–5(NP)	52.1	3	46	<0.05	42.2	4	42	<0.05
More attractive								
1(P)–3(P)	28.7	22	3	<0.05	12.3	10	5	ns
4(P)–6(P)	30.6	21	5	<0.05	8.5	3	7	ns
1(NP)–3(NP)	23.2	18	4	<0.05	19.3	18	3	<0.05
4(NP)–6(NP)	21.1	16	5	<0.05	20.2	10	12	ns

3(NP)] out of four comparisons in which the physical attractiveness of the imagined partner was increased was the number of females who changed in the 'more likely' direction significantly greater than the number who changed in the 'less likely' direction. In two comparisons [4(P)–6(P) and 4(NP)–6(NP)] more females actually changed in the 'less likely' than in the 'more likely' direction (though the difference was not statistically significant).

To analyse the importance of the chance to form a more durable relationship with the imagined partner, we compared responses to pairs of questions that differed only in whether or not such a chance existed. Table 8.4, which was constructed along the same lines as Table 8.3, summarizes these comparisons. These data indicate a marked sex difference: in every case but one [3(P)–6(P)], the chance to form a more durable relationship with the imagined partner caused a significantly greater number of females to change in the 'more likely' than in the 'less likely' direction; in no case, however, was this true for males.

8.4 DISCUSSION

Whether or not they already had a steady partner, and despite variations in the physical attractiveness of the imagined partner and the chance to form a more durable relationship with the imagined partner, males were consistently more likely than females to say that they would have sexual intercourse with an anonymous new person, although the risks of pregnancy, discovery and disease were, in this imaginary realm, absent. Certainly these data are not startling or counter-intuitive; on the contrary, they are precisely what every-day experience would have led one to expect. These data imply that male–female differences in willingness to engage in one-night stands and to have

Table 8.4

Effects of varying chance of forming a more durable relationship with imagined partner on response to questionnaire items

Response comparisons	Males				Females			
	% changed	Number more likely	Number less likely	p	% changed	Number more likely	Number less likely	p
1(P)–4(P)	27.6	12	12	ns	27.0	26	7	<0.05
2(P)–5(P)	25.3	16	6	ns	21.3	22	4	<0.05
3(P)–6(P)	29.4	11	14	ns	23.7	18	10	ns
1(NP)–4(NP)	34.7	20	13	ns	39.8	38	5	<0.05
2(NP)–5(NP)	33.0	19	12	ns	41.8	43	3	<0.05
3(NP)–6(NP)	27.4	15	11	ns	30.9	28	6	<0.05

sexual relations with new partners do not result merely from the different real-life risks males and females encounter. We have not, of course, conclusively demonstrated the existence of sexual dimorphism in one or more of the brain/mind mechanisms that underpin sexual desires, but our data are consistent with that interpretation. In short, these data support the view that men and women seem to differ in their tendencies to enjoy one-night stands and in the significance of partner variety for sexual attraction because men and women do differ.

We did not find, nor did we expect to find, that 100% of the males chose 'certainly would' and 100% of the females chose 'certainly not'. Human action, even imagined action, is determined by the simultaneous operation of many brain/mind mechanisms (most of which, presumably, are sexually monomorphic) and by the vagaries of individual life histories. Sex differences in responses to questionnaire items can only provide indirect clues about the natures of the sexually dimorphic mechanisms.

These data fit well with a number of recent studies that report no diminution in sex differences in sexual psychologies despite diminution in sex differences in sexual activities, despite the development of reliable contraception, despite the rise of the women's movement, and despite the increasing status and economic independence of many women (Sigusch and Schmidt, 1971; Houston, 1981; Wilson, 1981 Deaux and Hanna, 1984; Carroll *et al.*, 1985; Glass and Wright, 1985; Singer, 1985a and 1985b; Townsend, 1987). For example, Houston (1981) writes: 'The tendency for females to give more romantic responses [on a questionnaire] and males to give more erotic responses was expected and is in keeping with previous research. The suggestion is strong that the feminist movement and the generally observed societal shift away from male sexism has not altered significantly the eroticism–romanticism distinction, in spite of the sporadic romanticism of the males and eroticism of females'. And Carroll *et al.* (1985) conclude: 'The phenomenon of gender convergence in sexuality has been noted. Specifically, females have been catching up to males in incidence of premarital intercourse . . . It has been claimed that gender convergence is also occurring in 'interpersonal scripts' or motives, with a 'rhetoric of love' emerging in boys . . . If such a gender convergence in motives is occurring, it is not evident in these 1982 data.' Wilson (1981) compared male and female responses to the following question: 'With a new partner do you prefer to make love: (a) The first moment you can; (b) When you have got to know them a little better; (c) Not until there is some commitment to a steady relationship; (d) Not outside of marriage?' Males were far more likely than females to choose (a) or (b), and there was no significant difference between the responses of subjects under thirty and subjects over thirty; i.e., there was no significant trend toward gender convergence among the younger subjects. (We included Wilson's question in our questionnaire and found a dramatic sex difference in every socioeconomic, political, ethnic and racial sub-group.)

The data presented here also indicated that the physical attractiveness of an imagined partner is more important to males than it is to females, although decreasing physical attractiveness decreased erotic interest for both sexes. The fact that females were strongly influenced by decreases, but not by increases, in the imagined partner's physical attractiveness was not anticipated, and this matter ought to be addressed specifically in future research.

Our tentative interpretation of these data is as follows. Physical attractiveness is one determinant of sexual attractiveness for men and women alike, but it plays a much greater role in men's than in women's assessments (Buss and Barnes, 1986; Symons 1979). The males' responses to questionnaire items thus can be interpreted in a straightforward fashion, but the females' responses may indicate the existence of some sort of threshold effect: it may be important to women that a potential partner surpass some threshold of physical attractiveness, but increases in physical attractiveness beyond that threshold may be relatively unimportant.

This does not, however, account for the considerable number of women who indicated that they would be less likely to have sexual intercourse when the physical attractiveness of the imagined partner was increased. Only further research can illuminate this phenomenon, but here is one possible explanation. Other things being equal, women prefer more physically attractive sexual partners; but to some of the women in this study, other things are not likely to remain equal when male attractiveness is increased. Increasing the physical attractiveness of the imagined partner caused more women (both with and without steady partners) to change in the 'more likely' than in the 'less likely' direction only when there was no chance of forming a more durable relationship. When there was a chance of forming a more durable relationship, more women changed in the 'less likely' than in the 'more likely' direction. This may indicate that some women consider handsomeness a drawback in a potential mate, but not in a one-night stand, perhaps because handsome men have relatively abundant sexual opportunities and hence are perceived – probably with reason – as being relatively poor bets for sexual fidelity.

Our data also indicate a dramatic sex difference in the significance of the chance to form a more durable relationship with the imagined partner: such an opportunity consistently increased women's willingness to engage in what might turn out to be a one-night stand; but no such effect was apparent in men's responses. Although many men did change their responses when the 'chance to form a more durable relationship' was varied, in no case did the number of men changing in the 'more likely' direction significantly exceed the number changing in the 'less likely' direction.

Unfortunately, this male–female difference is impossible to interpret with confidence: it could indicate (a) that the men were less likely than the women to be favourably influenced by the chance to form a more durable relationship, (b) that the men were less attracted than the women by the specific characteristics we attributed to the potential partner, or (c) both. Although we attempted to give the potential partner characteristics that both sexes would find appealing, we now suspect that these characteristics were more likely to appeal to the women than to the men in our sample. For example, an architect is almost certain to be older than most of our subjects, and this might well have affected men and women differently. Future questionnaire items designed to investigate this matter probably should attribute less specific characteristics to the potential partner than our questions did, and these characteristics should be ones that have been shown empirically to appeal equally to males and females (Buss and Barnes, 1986). We predict that responses to more sophisticated questionnaire items will reveal that although men and women may be equally interested in establishing durable relationships with desirable members of the other sex, the opportunity to do so will not significantly affect men's willingness to have sexual relations with anonymous new partners.

We have interpreted our data as supporting the hypothesis that one or more of the brain/mind mechanisms that underpin human sexual psychology is sexually dimorphic. Social and behavioural scientists, at least in the US, usually treat this hypothesis as if it were somehow extraordinary (as if it were, for example, analogous to the hypothesis that spoons can be bent by the unaided power of the mind) and hence required extraordinary supporting evidence in order to be taken seriously. Although social and behavioural scientists rarely explicitly mention the human brain/mind at all in their explanations of male–female differences in sexuality, these explanations, which normally invoke society, culture, scripts, roles, social learning and so forth, imply a definite view of the human brain/mind. Specifically, such explanations imply that however males and females come to differ in their sexual desires and dispositions, they do so with identical brain/mind mechanisms. In other words, to explain male–female differences in terms of culture, society, scripts, roles, or social learning is, essentially, to imply that, in the absence of conclusive laboratory evidence to the contrary, it is reasonable, prudent, and parsimonious to assume that the human brain/mind is sexually monomorphic.

Evolution-mindedness, however, implies that precisely the opposite assumption is the reasonable, prudent and parsimonious one: to someone whose thinking is informed by evolutionary biology and the data on non-human animal behaviour and brain dimorphism, the likelihood of the human brain/mind being sexually monomorphic is, for all intents and purposes, nil. In this view of life, it is the hypothesis that men and women are fundamentally alike in their sexualities that is extraordinary and that requires extraordinary evidence in order to be taken seriously.

144

We hope that future questionnaires will be designed not merely to investigate sex differences that are already reasonably well conceptualized (and that most men and women in the street already take for granted) but to augment and refine the conceptions themselves. One strategy, of course, would be to give the same questionnaire to samples that vary in age, social class, nationality, and so forth. (A serious limitation of the present study is the relative youth and inexperience of the subjects; it will be interesting to compare their responses with those of an older, more experienced sample.)

A less traditional strategy, however, would be to progressively alter questionnaire items with the goal of maximizing sex differences in responses. If a question that had been designed to reveal a sex difference failed to do so, instead of concluding that men and women must be identical in this respect, the investigator could experiment with variants of the question to see if any would elicit a sex difference. If a variant did turn out to elicit a small sex difference, the investigator could then write variants of this question to see if this difference could be magnified. In short, questionnaires could be used not just to test hypotheses but as part of the process of hypothesis formation itself. Comparing questions that did not elicit a sex difference with variants that did, and questions that elicited minor sex differences with variants that elicited major ones, might eventually expand and refine our conceptions of male and female natures.

8.5 SUMMARY

Four hundred and fifteen students (232 females, 183 males) enrolled in introductory-level general education courses completed an anonymous paper-and-pencil survey on various aspects of sexuality. Males were consistently more likely than females to say that they would have sexual intercourse with an anonymous new person, even though the subjects had been asked to imagine that such intercourse entailed no risk of pregnancy, disease or discovery. Male sexual interest in an anonymous new partner generally increased when the imagined partner's physical attractiveness was increased, decreased when the imagined partner's physical attractiveness was decreased, and was not consistently affected by the chance to form a more durable relationship with the imagined partner. Female sexual interest in an anonymous new partner generally decreased when the physical attractiveness of the imagined partner was decreased, increased when there was a chance to form a more durable relationship with the imagined partner, and was not affected when the imagined partner's physical attractiveness was increased. These results provide some support for the hypothesis that the human brain/mind is sexually dimorphic in one or more of the mechanisms that underpin sexual experience.

ACKNOWLEDGEMENTS

We thank David Abbott, Christine Hulihan and Lars Perner for their assistance with quantitative and statistical matters and D.E. Brown and Kelly Hardesty-Ellis for their helpful comments on an earlier draft of this chapter.

REFERENCES

Buss, D.M. and Barnes, M. (1986) Preferences in human mate selection. *J. Pers. Soc. Psych.*, **50**, 559–70.

Carroll, J.C., Volk, K.D. and Hyde, J.S. (1985) Differences in motives for engaging in sexual intercourse. *Arch. Sex. Behav.*, **14**, 131–9.

Deaux, K. and Hanna, R. (1984) Courtship in the personals column: the influence of gender and sexual orientation. *Sex Roles*, **11**, 363–75.

Gibbons, J.D. (1976) *Nonparametric Methods for Quantitative Analysis.* American Sciences Press. Columbus, Ohio.

Glass, S.P. and Wright, T.L. (1985) Sex differences in type of extramarital involvement and marital dissatisfaction. *Sex Roles*, **12**, 1101–13.

Houston, L.N. (1981) Romanticism and eroticism among black and white college students. *Adolescence*, **16**, 263–72.

Sigusch, V. and Schmidt, G. (1971) Lower-class sexuality: some emotional and social aspects in West German males and females. *Arch. Sex. Behav.*, **1**, 29–44.

Singer, B. (1985a) A comparison of evolutionary and environmental theories of erotic response Part I: Structural features. *Journ. Sex Res.*, **21**, 229–57.

Singer, B. (1985b) A comparison of evolutionary and environmental theories of erotic response Part II: Empirical arenas. *Journ. Sex Res.*, **21**, 345–74.

Symons, D. (1979) *The Evolution of Human Sexuality.* Oxford University Press, New York.

Symons, D. (1980) The evolution of human sexuality revisted. *Behav. Brain Sci.*, **3**, 203–14.

Symons, D. (1987a) Can Darwin's view of life shed light on human sexuality? In: *Theories of Human Sexuality* (eds J.H. Geer and W.T. O'Donohue), Plenum, New York, pp. 91–125.

Symons, D. (1987b) If we're all Darwinians, what's the fuss about? In: *Sociobiology and Psychology: Ideas, Issues and Applications* (eds C. Crawford, M. Smith and D. Krebs), Lawrence Erlbaum Assoc., Hillsdale, N.J., pp. 121–46.

Townsend, J.M. (1987) Sex differences in sexuality among medical students: effects of increasing socioeconomic status. *Arch. Sex. Behav.*, **16**, 425–41.

Trivers, R.L. (1972) Parental investment and sexual selection. In: *Sexual Selection and the Descent of Man 1871–1971* (ed. B. Campbell), Aldine, Chicago, pp. 136–79.

Wilson, G.D. (1981) Cross-generational stability of gender differences in sexuality. *Person. Individ. Diff.*, **2**, 254–7.

Wilson, G.D. (1987) Male–female differences in sexual activity, enjoyment and fantasies. *Person. Individ. Diff.*, **8**, 125–7.

CHAPTER NINE

Human courtship behaviour: biological basis and cognitive processing

Karl Grammer

9.1 INTRODUCTION

In 1975 Kendon described the literature on human courtship as almost non-existent. Almost ten years later, Hinde (1984) found our knowledge still fragmentary, and, moreover, in need of an integrative framework. Indeed, most of what we know about courtship originates from just a handful of direct observations. What remains consists of interviews carried out at different stages of courtship. There are at least two reasons for this situation. First, it seems to be difficult to obtain sufficient and convincing data, despite information gathered by questionnaires. Second, cross-cultural comparisons seem to show that the behavioural variability is high. Moreover, courtship seems to have undergone historical change (Cook, 1981), even though, as a result of biological restraints, courtship behaviour appears to be a bastion for the strict performance of stereotyped gender role behaviour. In this chapter I consider existing data and develop empirically testable hypotheses in the light of socio-biological and cognitive theory. In this way I hope to focus on the study of this important aspect of human life.

9.2 BIOLOGICAL THESES: CONFLICT OF INTEREST AND THE BATTLE OF THE SEXES

From a biological point of view, the ultimate function of courtship is the maximation/optimization of the reproductive success of the individuals involved. Genotypic and phenotypic traits which guarantee this success are distributed unequally in a population (i.e. mates differ in quality). Thus the selection of an appropriate mate will become the main theme of courtship.

147

This first requirement generates sexual attraction for partners whose characteristics, when transmitted to their offspring, will increase the reproductive success of the latter. These basic conditions should lead to intrasexual competition for mates in both sexes.

Intrasexual competition then would accentuate advertisement for those phenotypic traits signalling mate quality. Thus males, and to a certain degree females (Maynard Smith, 1974), should compete through sexual advertisement. The greater the amount of advertisement an individual performs, the greater its relative number of possible choices should be. On the other hand, advertisement means costs in time and energy and enhances the chance of detection by predators. This forces the level of advertisement to a competitive optimum. An individual has to display just slightly more than the others do, as long as this is not detrimental to his/her own survival (Parker, 1983). This process should control the intensity of advertisement.

Another assumption is that costs for mating are different for the sexes:

Females Due to internal fertilization, costs to females are higher than to males. The early survival of the offspring depends mainly on maternal care. If an infant does not thrive, a female will have to invest more than a male to bring a second infant to the same stage (Dawkins and Carlisle, 1976). Thus, for the female, male assistance reduces her costs for mating. As a result, a female should show greater interest in maintaining a stable pair-bond. Moreover, if paternal care plays an outstanding role for infant survival, males are more likely to vary in quality. Selecting a 'bad' mate could endanger the survival of the female's rare offspring. In consequence, females should be more choosy than males (Trivers, 1972).

Males In contrast to females, males have lesser costs. They could thus try to augment their reproductive success by philandering. This, however, is only possible if the effect of paternal care is negligible for offspring survival. As soon as paternal care is indispensable, the male faces a problem. In contrast to the female, he cannot be sure that the offspring he is caring for is his own Possible philandering by the female would cause fear of cuckoldry in the male and would also threaten his investment.

Consequently we find an influence of investment on the intensity of advertisement. Direct intrasexual competition will be most intense in the sex that invests least in a given pairing (Darwin, 1871; Bateman, 1948; Trivers, 1972). Therefore it will be the males who experience most intense direct competition. Thus males should tend to more overt advertisement. Also, if inter-male competition occurs, male choice and variance in mate quality should create an effect called assortative mating. This means that pairings should show significant positive correlations for mate quality.

Note that all these considerations hold only for the social situation where paternal care plays an important role for offspring survival. How far this holds for human evolutionary history may be answered only speculatively.

Nevertheless, we assume that paternal care (under given ecological conditions) does constitute a necessity for the rate of offspring survival.

This theoretical assumption about intrasexual competition and intersexual conflict leads to the following ultimate considerations (Hinde, 1984). Although females should compete for males, they should be choosey. The males should compete vigorously and their selective level should be lower than that of the females. The males should be attracted to females seen as receptive and whose characteristics indicate that they would rear the male's offspring successfully. In contrast, females should be attracted, in the first place, to males on the basis of their prospective investment.

Males, then, should have an aversion to invest in relationships with females who are sexually promiscuous. Males should invest only in relationships where they are unlikely to be cuckolded by a female. Females should have a tendency to avoid exploitation and should try to test out the males' willingness to invest in the relationship.

Sex differences in perception of optimal mate qualities should cause sex differences in the tactics and the quality of advertisement. Males should advertise in a manner congruent to female perception of optimal males, and vice versa. We should at least expect different strategies which can be used to prevent exploitation.

So far we have drawn a static picture of mate choice, which surely does not consist of a single decision but is a process of negotiation between the partners. The decision to approach could be triggered by the attractiveness of the partner. After an initial decision to approach is made, the course of courtship should be guided by the assumed investments of the prospective partners. Theoretically, we should expect a decision point (to stay or leave) to occur earlier for the females. This is an effect of their higher investment in the offspring.

These ultimate considerations allow us to state precise hypotheses on human mate selection. From these we can make predictions on the tactics that may be used in finding a mate, and in deciding if he/she is an appropriate mate. Finally we can predict that the ego will try to convince the prospective mate that he himself is appropriate for her, or vice versa.

9.3 THE COGNITIVE BASIS OF DECISION PROCESSES IN COURTSHIP

The process of finding, deciding upon and establishing a courtship relationship may be viewed as an attempt at social problem solving. The intentions or goals of one partner may not be shared by the other. If successful mating is to occur, the goals of the partners have to converge. The question here is, which tactics are efficient in achieving the pursued objectives? There have been numerous theories in social psychology and recently in sociobiology stating that the choice of an action depends on the costs of the action and its prospective benefits. Homans (1961) and, in a comparable approach, Thibaut and Kelly (1959),emphasized that the choice of an action depends on a plan which is evaluated in terms of the outcome produced. Each individual tries to maximize this outcome, and tries to establish at least equity, i.e. equalize cost and benefits with its partners. The problem with an empirical approach to such theories arises from the attempt to measure value, outcome, benefit.

Another approach, generated in linguistic theory, and derived from cross-cultural research on verbal requests (Brown and Levinson, 1978) concentrates on the central terms of risk. This theory relies on the fact that people are quite capable of judging they can reach a defined objective. Risk, then, would describe the assessment of the possibility of reaching a certain goal. Thus risk describes the possibility of non-compliance by the partner.

Another individual might not even share the goals of the actor and interrupt the actor's ongoing behaviour by constructing 'behavioural blocks' (Charlesworth, 1978). These blocks have to be removed by the acting individual in such a way that he/she is able to reach the original goals. The removal of blocks is, according to Charlesworth, the adaptive function of intelligence. But 'social problem solving' is not only the art of removing existing and obvious interruptions in the behavioural stream. It appears to be less time and energy-consuming to make a plan which reduces the likelihood that a behavioural block occurs, than trying to remove existing blocks. An individual should thus weigh the existing behavioural alternative. Then the individual is able to apply a suitable strategy which lowers the probability that the goal reaching attempt will be blocked.

In order to make a plan, an individual has to assess the possible risk. This procedure has to start with an evaluation of the goal under quest. Each possible goal seems to have a defined risk. The knowledge regarding risk and goals is shared by all members of a group. Risk can certainly be modified through cultural rules or norms. It can also be changed by motivational factors, as for instance the high attractivity of a goal.

A further assessment of risk requires information that allows predictions to be made on the possible reactions of the target person. Thus information gathering is the most prominent feature in interactions. But the gathered information has to be processed and compared to stored information. This can be done when qualities of interactions form constant patterns over time. If regularities emerge in patterns, they are abstracted as concepts. These concepts form the basis for the deduction of future events. One of the most prominent features of concepts is that they tend to create dichotomies. Thus we should expect at least two axes in their organization: friendly–hostile and dominant–submissive are the most prominent ones. Conceptualized as relationships they allow a prediction of the behaviour a target person will show in a wide range of situations. This process also works among strangers.

Humans tend to show signs of dominance and social attitudes in their appearance and behavioural style. They tend to constantly rank and classify prospective interaction partners (Zetterberg, 1966). Dominance and social distance are negotiated and clarified even more in the opening of interactions (Eibl-Eibesfeldt, 1984).

We assume a constant, additive relation between the above mentioned parameters, as suggested by Brown and Levinson (1978). If high status and high social distance on the side of the actor are predictors of the possible compliance of the target person, the perception of high status and high social distance should create low or high risk for the goal pursued. Targets and goals thus have a risk feature which we can describe as the external organization of the social problem-solving attempt (Grammer, 1982). External organization provides the framework for the sequence, as well as for the quantity and quality of tactics of a goal reaching attempt. The question is, therefore, which tactics suit which risk conditions?

Before starting to act, the individual has to make a decision. This could be done by comparing the risk with the necessary costs and the possible benefits. If the costs in meeting the risk are higher than the estimated benefits, an individual will probably not initiate any behaviour at all. Thus we should expect the following situation: if an approach takes place, an initial encounter will at least last for a test-phase which allows the gathering of additional information.

Internal organization circumscribes the structure of a strategy selected out of a pathway network which includes all possible ways to a goal. A prerequisite for the development of strategies is the existence of tactics which enable the acting person to produce predictable effects (behavioural changes). Free variation in the use of tactics, however, is restricted by a number of constraints. The target person always interprets behaviour causally: he/she will interpret the actor's behaviour as directed towards a goal, i.e. as a means used by the partner to maximize the outcome of his/her behaviour. The actor is able to overcome this tendency by constructing behavioural 'detours'. If he/she chooses tactics which prepare the grounds for reaching the goal, then the target may not even realize that he/she has become compliant. Chisholm (1976) discusses this problem in terms of good and bad moves. The main feature of a good move would be that it does not restrict the actor's possibilities of further action. In contrast, a bad move would restrict the actor's possibilities or even interrupt the goal-reaching attempt. Tactics thus have to have a risk-dependent escalative potential. On the other hand, the limited amount of time available puts a certain restraint on the use of detours. The amount of time can be further reduced by competition. The time limit forces the actor to clarify his intentions within a certain time frame. Thus a dilemma arises out of the necessity to construct detours and, at the same time, to clarify intentions. This dilemma defines the qualities and the content of the

151

tactics and determines their sequential use. Thus not only cost and possible benefits trigger the approach; the individual also has to include a comparison of the time limit and his/her available tactics.

If the quantity of risk is directly correlated with compliance of the target person, then low risk allows a direct approach. On the other hand, high risk demands detours, in order to increase the possibility of reaching the goal pursued. In this case the intentions are revealed step by step.

Empirical evidence suggests that the above mentioned hypotheses are useful for the description of, and for predictions on the structure of strategies in all situations where goals have to converge between interactants. Thus we might talk of a universal 'social-problems-solving algorithm'. This algorithm describes the nature of strategies and the risk-dependent quality of possible tactics. The basic ideas were supplied by Grammer (1982, 1985) for the organization of strategies of intervention in conflicts among pre-school children and by Grammer and Shibasaka (1985) for access to play groups.

9.4 HUMAN COURTSHIP: GOAL DIRECTED ACTION

The approach proposed depends mainly on the possibilities of observing and classifying 'goals'. For a definition we will follow Cranach *et al.* (1980), who define goal as an imagined, aimed at stage at the end of an action. According to Cranach *et al.*, goals are structured both in time and hierarchically. Goals are describable as higher order goals with their respective subgoals. We have to take into account, however, that there is constant interaction between the different levels of goals. Another complication is the fact that goals may be adapted to situational changes during interaction.

In courtship behaviour, we find a broad spectrum of goals, but the ultimate goal of maximal/optimal reproduction is of prime importance. As a logical consequence we find sexual intercourse on the next level. At the same level, wishes for the establishment of relationships of different qualities will play a role. These are indeed the main goals found by Kirkendall (1961) in interviews with 200 American college students. It is on this level that we first find differences between the sexes. Owen (1982) analysed advertisements in a 'lonely hearts column'. In the analysed advertisements males seem to look for partners for 'fun times' (McDaniel, 1969), whereas females looked for permanent relationships and marriage. But this level may be complicated by the existence of goals not directly bound to reproduction. Self-esteem, achievement, approval or maintenance of power or status (mainly males) and even material or physical exploration are not uncommon (Skipper and Nass, 1966).

All three types of goals have the same subgoal: the choice of a target person of the other sex. This goal demands presentation as potential mate. The first second-order subgoal is identifying sex. According to Skrzipek (1981, 1982) and Horvath (1979) people seem to have a sex-specific template for the recognition of the other sex. The relationship between shoulders and waist in males (broad shoulders, small waist) and the hip-waist relation in females (small waist, large hips) is used, among other clues, to judge gender. This seems to be a universal phenomenon, common to all cultures. At the same level of goals we find a necessity to identify the reproductive condition of a potential partner.

These goals are followed in time by attracting specific attention, as a first step for making contacts. Overcoming aggression may be crucial at this point. In a study by Sack *et al.* (1982) one out of every four college students (independent of sex) reported that he/she had been either a victim of violence or had engaged in some form of violence in dating situations.

After making contacts the evaluation of the partner is necessary. Females should be interested in information on the males' potential for protecting and providing for her and her offspring whereas males should evaluate the females' tendency for cuckoldry. But there are, as far as we know, no data on the actual evaluation process. A hint may be given by Davis (1978) who found that males proceed faster to intimate topics than females in an acquaintance exercise.

At this point the female should pursue a second goal: testing out the male's willingness for investment. Again there are no data on the actual process or on the representation of this goal. The same holds for the next goals: achieving coordination (Barash, 1977) and maintaining spatial proximity. Holding attention, sexual enticement and, finally and, according to the theories probably more important for females: avoiding exploitation. Berk reports in his observations of a singles' dance that this is a goal women try to achieve by creating female–female alliances. This is underlined by the fact that atrocity stories are an important topic in conversations amongst the women (1977).

In a review of apparent goals, we find indications of a hierarchy and some evidence for biologically based goal-structures. Females should weigh goals differently than males, and we should also expect different risk assessments for the goals. The quality of the relationship pursued may play an important role. High expectations on quality of relationships, as with marriage intentions, could augment risk for the acting person.

9.5 STRATEGIES, TACTICS AND RISK PERCEPTION

In the discussion of actual behaviour, we will follow the time structure of the goals. We will also look for the influence of the higher-order goals. One caveat should be kept in mind: the fast-growing literature refers mainly to American college populations.

We also have to note that free decision and courting is not the usual case. Mates in traditional cultures are often selected by parents or relatives for socio-economic reasons, and courtship is often carried out as a complex ritual.

Cultural influence on decisions finds its expression in regulation through kinship, social class, and often by the age of the potential partners.

9.5.1 Advertisement and attracting specific attention

Attracting attention is the first goal. Attention is elicited through the display of signals that excite the interest of possible mates. No doubt these include a person's physical looks, clothing and behavioural style as a basis for the decision to approach him or her.

Observations and interviews indicate that non-verbal solicitation is mainly done by the female. Scheflen (1965) observed many non-verbal cues demonstrating courtship readiness. Symonds, in her observations of group-sex parties (1972), found the same signs, which she calls 'non-verbal come-ons'. The starting point is eye-contact followed by immediately looking away or lowering of the eyes. The next step is lowering or turning away of the head which is followed by mutual eye-contact again. More direct signs are: not looking away (outstaring), fixing the target person and starting to breathe synchronously. If the other person notices the contact, the eyes wander up and down his or her body. In addition Symonds describes typical patterns of sitting, standing (flexion of muscles or hand on the hip) and walking (flip of the hip). According to Givens (1978) the recognition phase is marked by head cocking, pouting, primping, eyebrow flashing and smiling. Moore (1985) who followed single females through a discotheque, indicates that it is the female who determines and controls the approaches of males by exhibiting or withholding displays: 'They can elicit a high number of male approaches, allowing them to choose from a number of available men, or they may direct solicitations at a particular man'. According to Moore, the number of male approaches correlates directly with the amount of female solicitation. Perper and Fox (1980) observed in bars that women often make the first but subtle move, usually little more than standing close to their target. Finally, Cary (1976) found that conversations only started when the female glanced at the man at the beginning of the encounter.

Women seem to be exquisitely familiar with what occurs during flirtation; on the contrary men are quite ignorant. Women can describe in great detail how they and other women flirt and pick up men. Even quite successful men seem to have no idea how they attract women and what happens during flirtation (Perper and Fox, 1980). 'I just knew that it would work out' was an answer often recorded by Kirkendall (1961). According to him, males' decisions are based on the reputation of the girl, her clothes, and the place where they found her (bars, etc.). Berk (1977) in his observations of singles' dances found that men enhanced presentation by typical arrival and departure patterns: coming with a friend, coming late and leaving early were widespread.

9.5.2 Target choice

Besides female solicitation, other factors influence the decision to approach. Regardless of context, physical attractiveness has been found to be the primary basis on which dating selections are made. This is the case both in terms of what people say they want, and in choices actually made. Coombs and Kenkel (1966) pointed out that physical attraction was the most important variable for both men and women, but more influential for men. Men were attracted to women who shared their sexual attitudes. Women's choices were modified by race, religion, intelligence, campus status of men and by concern about dancing ability and dress. Walster *et al.* (1966) added the factors of personality and popularity and called the whole complex 'social desirability'.

The status of the male, his economic attributes and thus his abilities to offer financial security influence the female's decision (Rubin, 1973; Harrison and Saeed, 1977). In a non-college population, females preferred higher status males over low status males (Green *et al.*, 1984). Furthermore elder males and younger females had an enhanced dating potential in this and other studies (Folkes, 1982). Sigall and Landy (1973) make corresponding observations. People usually ascribe higher status to men who are accompanied by attractive women. Women however, do not only seek financial security in males. Their value as a social partner also correlates with the magnitude of emotional security a male can provide (Fowler, 1978). Finally, for the overall weighting of traits, Coombs and Kenkel (1966) found, in general, that females have higher aspirations for partners than men.

The basis of the main trait influencing decision, physical attractiveness, is not yet clear. Clarification of this point would be essential for the inference that higher attractiveness would guarantee higher fitness in the offspring. The assessment of physical attractiveness seems to be based on a number of traits rather than any single one. Eibl-Eibesfeldt (1984) summarizes attractiveness factors: regularity of features, smoothness of complexion, optimum stature and good physique. Thus initial attraction is based largely on characteristics which are not unrelated to health condition or sexual potential.

The size of breasts and buttocks correlate with sexual attractiveness, although this correlation is culture-dependent (Hess, 1975). Cant (1981) hypothesizes on the basis of Frisch's (1975) statement 'fat is the issue', that big breasts and buttocks signal a female's potential for parental investment in her offspring. Critical fat levels seem to be responsible for menarche and thus ovulation (showing receptivity) and for lactation (showing possible female investment). It is possible that breast size is an indicator of fertility and potential parental investment, but in view of this it is interesting that breast size influences sexual attractiveness and not mate selection.

Numerous authors stress that attractiveness seems to be dependent on cultural norms (Berscheid and Walster, 1974). But cross-cultural comparison of mate selection (from the literature) is almost impossible at present because information, where it exists, is not very reliable and seldom complete (Rosenblatt and Anderson, 1981). This is an open field for further research.

The same holds for the expected influences of social class. According to Eckland (1982) it is still an unsolved problem whether social class or spatial proximity play a role in mate selection. Clarke (1952) found that in Columbus, Ohio more than half of the married couples under study lived within walking distance of each other at the time of first dating. Humans do not live randomly dispersed, however, and in western society, neighbourhoods are normally inhabited by the same social class. We are not yet able to decide whether attempts to maintain spatial proximity, or social origin itself, are factors for mate-choice.

We would assume that a decision to approach is made when an individual is sighted whose traits are at their maximum value. Huston (1973) tested the selection of possible mates as dependent on their attractiveness. He started with the assumption that males would prefer only highly attractive females. If he gave men the information that they would always be accepted, men chose highly attractive females. If he told them it was not sure that the females would accept them, they chose partners slightly above medium attractiveness. Males thus perceived their chance of being accepted as dependent upon attractiveness of the female. Murstein (1976) suggests that individuals try to match their physical appearance, one with the other, and choose partners whose physical appearance is comparable in attractiveness. He also presented data to support this theory using independent ratings of the photographs of married and courting couples. The results suggest that couples are indeed more similar in physical attractiveness than would be predicted by chance alone.

The picture of the traits used for decisions in mate selection and the actual behaviour in the opening phase appear to reflect the biological hypotheses. Females seem to be choosey, looking for high status males older than they are themselves; males who can provide material and emotional security in a relationship. In contrast, males decide on the basis of sexual and physical attraction and prefer younger females. To date, however, there is no proof that these factors are related to fitness, i.e. that these traits guarantee more offspring.

Where individual variability for the above-mentioned traits is present, we should expect that males and females can be ranked according to those traits. If maximum reproductive success is coupled with maximum occurrence of traits, then competition arises. If all men like beautiful women, and beautiful women seek emotionally secure men of high status, then beauty and dominance will be selected for. As an outcome of competition, though, we find actual mate selection which is based on resemblance in one or more characteristics. Usually, in humans, assortative mating is positive (i.e. greater than chance similarity of mates). No one has yet demonstrated negative assortative mating for any human trait in any large population in a statistically significant way. Tharp (1963) reviewed the following traits: race, ethnic origin, social class, age, religion, education, intelligence, various personality traits, physical characteristics (height, weight, complexion), values, interests, residential propinquity and many other variables. Intelligence quotient, age, and formal education showed the highest degree of assortative mating in European and North American Caucasoid populations.

The concordance of traits in mate selection might be due to risk-perception. An approach could be triggered by the comparison of risk, the available tactics and a rating of one's own attractiveness. Whether an individual starts acting depends on whether he/she can handle the amount of risk assumed to be present in the situation.

Furthermore, if he/she takes his own attractiveness into account, he/she ends with a partner of the same value. If we add the time-limit which is created by male–male competition, we can hypothesize that males should be more aware of their own attractiveness to females than vice versa.

9.5.3 Making contacts

According to McCormick and Jesser (1983), the man is not the sexual aggressor eagerly pressing himself on the coy and reluctant women. We saw that the first moves on the woman's side may be subtle. Thus it seems understandable that men come to believe that they started the interaction themselves. And indeed, in most of the cases, the first overt moves are on the male side. If males are in competition with other males, they must take their chance as soon as they see risk lowered through solicitation. Kirkendall's interviews (1961) show that males usually start indirectly with suggestions and invitations. The majority of males state that it is easier for them to start with indirect or conventional speech acts, only implying interest indirectly. There is thus no relevant statement for the specific moment. The content of the verbal utterances does not endanger the arising relationship. Interestingly enough, most of the students stated that men should take the initiative for making contact. Observations on verbal invitations made by Symonds (1972) in 'swinger parties' and by Roebuck and Spray (1970) in a cocktail lounge underline this finding. Symonds summarizes: 'I am willing to generalize, that with the male propositioner the directness of the proposition is positively correlated with his perception of acceptance'. According to her, an indirect proposition would be the best for both partners, because this way the propositioner does not feel 'put down' through rejection and the propositionee is not committed. Cook (1981) concludes, on the basis of cultural comparisons made by Ford and Beach (1952), that direct invitations seem to be an exception. He describes most of the invitations as vague, symbolic or non-verbal. The point about direct verbal invitations is that they tend to require direct verbal answers, which either commit the speaker or offend the asker. Apart from this, we also find direct, unmistakable invitations, in contrast to the overall rule of indirectness. The reason for these could be found in risk perception. If risk plays a role, first utterances should vary in risk, depending on their directness. Indeed this is a critical point, because here rejection can occur for the first time. This is underlined by Berk (1977) who describes males' strategies for the management of rejection. Common tactics are: denial (do not look at it, it does not exist), re-definition (talk it away), enhancing presentations, limiting involvement, putting others down (I'm better than they are) or withdrawal.

9.5.4 Evaluation and achieving of co-ordination

First we will look at the behavioural repertoire which is used during the first stages of courtship.

Givens (1978) describes a high degree of female ambivalence: primping, object caressing and glancing at and then away from the male. Females and males in conversation appear highly animated and Moore (1985) found that women were highly excited while talking to men. They laughed, smiled and gesticulated frequently after making contact. This is the point where the woman reaffirms her interest: by nodding often, leaning close to the man, smiling and laughing at higher frequencies.

All of these descriptions have two points in common. They portray non-verbal behaviour which also occurs in normal interactions. In courtship, however, the frequency of performance is said to be higher. In addition, quality of the behaviour appears to be different. Descriptions such as: somewhat longer, transient, somewhat more than usual, vehement and quick are qualitative markers for this. Non-verbal behaviour in courtship holds an element of exaggeration which is now directed at a defined target person.

Achieving coordination means convincing the partner that some common base can be established. This can be reached by establishing common ground in the first utterances. Common ground is established non-verbally by the female through nodding (Morris, 1978) or by 'lean', a very common solicitation pattern. 'Lean' is performed by moving the upper part of the body in the direction of the man and thus demonstrating unison (Moore, 1985). Perper and Fox (1980) found that they could predict by the degree of non-verbal synchronization whether a couple left a bar together or would separate again. Dancing together plays a prominent role in courtship. Dancing is a test phase which makes it possible to evaluate whether a partner is able to share a common basis. Furthermore, the amount of non-verbal synchronization might signal readiness.

158

The Kirkendall (1961) interviews are the best source for verbal strategies. An 18-year old male reports: 'I had it all down to a pretty good science by that time. I could take her ideas and make them appear that we both agreed on the same thing'. Some of the students developed sophisticated systems of verbal persuasion. Other students followed the motto: 'if you start to talk, you'll talk yourself right out of it'. Verbal persuasion by males often refers to appeals based on females' desires. These appeals are often indirect and ambiguous. We find appeals to wishes for love and a permanent relationship, to the wish to appear intelligent or even fair. Finally there are attempts to diminish or evoke anxiety through threats of different kinds. Self-disclosure, i.e. the voluntary revelation of emotional cues or information on the self, is another strategy. Berk (1977) describes the 'sad tales tactic' which utilizes appeals to compassion.

These findings apply to the males' tactics, which seem to take the active part at this stage, whereas the female controls the situation by her non-verbal behaviour. This situation led Broverman *et al.* (1972) to the assumption that in the early stages of relationships men show more structuring activities. This was verified by Davis (1978): in a role play men were the architects of the encounter. They chose the topics independently from the females, which was not appreciated by the latter. The males' pleasure was the higher, the more self-disclosure occurred in the females. In contrast to these findings Kendon (1975) found in his analysis of a 'kissing-round' that the active behaviour of the man was controlled by the presence or absence of non-verbal signs in the female. Vivid facial expressions shown by the women regulated and modulated the approach and orientation of the man. If the woman smiled with lips closed, he started kissing her, if she smiled showing upper teeth, he looked away, this being another hint of female non-verbal control.

It is a saying that the harder women are to get, the more attractive they are to men. Walster *et al.* (1973) could not prove this hypothesis. On the contrary, men appeared repelled by such females. But what was found was that a woman could augment her desirability by attaining such a reputation and then making clear to the present partner that he was the only exception, thus demonstrating selectivity.

What about self-presentation on the male's side? Berk (1977) found that men foster positive impressions by demonstrating coolness, showing that they lead an exciting life, and by claiming prestigious occupations. In addition, it is of advantage if a third party, usually a friend, can verify these statements. Thus the friend is able to testify credibility and respectability which, in return, encourages females to trust the male. Kirkendall's males tried to impress their females by showing them how many friends they had, by the clothes they were wearing or by demonstrating their capabilities.

A further point would be to avoid exploitation on the females' side. This could be done by forming alliances with other women. Berk (1977) observed that women who came to the dance alone quickly formed alliances with a female friend. The allies found thus quickly exchanged information on the exploitative potential of males.

In the further course of courtship we find that argumentative–persuasive communication diminishes. Appeals by the males to the self-esteem and self-determination of the women now become more important. Further tactical moves are the establishment of mutual regard and attempts to appear or to be actually predictable in one's behaviour. The female usually presses for commitment at this point. If she does this too emphatically the male feels threatened and loses respect (Kirkendall, 1961).

The course of courtship (for the observed college populations), so far, is structured on quite different levels. Non-verbal signs emitted by the female show acceptance or refusal. These signs indicate the course of risk development for the male. On the verbal level, males try to avoid directness. In western societies, the males appear to have relative autonomy and the opening initiative is on their side. Thus self-presentation of the male on this level seems to be the important issue. Self-presentation is a strategic means by which risk is reduced for the next decision. Within self-presentation, males display those characteristics the female seeks in an optimal partner.

With the next goal, we find a distinctly marked risk threshold. The first stage is attaining physical contact. Most of this is done indirectly: touches appear accidental (Symonds, 1972). Moore (1985) observed a female tactic she calls breast touch or brush which is a short contact of the female's breast with the male's body. It was difficult to tell, except by length of time of contact, whether or not the movement was purposeful. Here again we find indirect approach. In a study by McCormick (1979) the answers given by students, when questioned on a hypothetical sexual encounter, indicated that both males and females prefer indirect strategies. They state for instance: 'I would test my limits by holding hands, sitting closer to the person.' But on the other hand, a highly direct arousal strategy is also popular with both males and females. Male: 'If she gives me the come on, then I would proceed very vigorously'. This occurs slightly differently on the female's side: 'I would try to be very sexy . . . a few sighs here and there . . . this would probably be all . . . aside from wearing something slinky and bare'. Attempts by males to persuade females to sexual intercourse often employ indirect verbal-strategies in order to avoid simple yes/no questions. Male: 'I never ask a girl if I may unbutton her blouse. I ask if it unbuttons or unsnaps, or if it unbuttons in the back or in the front. This way she has less chance to say no' (Kirkendall, 1961).

So far, it seems that sexual enticement is consistent with the gender role stereotype, although we find again a high degree of non-verbal control on the female's side. But the problem seems to lie in the analytical approach. Most of the results are generated out of interviews, thus relying on perception and consciously traceable information. It could be that gender roles are also generated on a perceptual level. This is outlined by the fact that men usually say they would use seduction significantly more often than women (McCormick and Jesser, 1983), a fact clearly contradicted by the non-verbal solicitation results.

However, (McCormick and Jesser, 1983) both partners prefer seduction over all other strategies of making a potential partner agree to sexual intercourse. When describing their personal use of power via various strategies, men use power related strategies significantly more often than women do in order to achieve sexual intercourse. Women use power-related strategies significantly more often than men do in avoiding sexual intercourse. Aggression, as for instance verbal threat, is widespread among males (Kirkendall, 1961). Males also report, however, that females started the aggression.

9.6 THE COURTSHIP GAME: COGNITION AND DETERMINATION

9.6.1 Risk-estimation: prerequisite for mate selection

Risk estimation seems to play a considerable role in courtship. For males, at least, high physical attractiveness in the female creates high risks of rejection. Thus, ultimate necessities of male choice, if high physical attractiveness marks a good partner for maximal rearing of offspring, are modified by risk perception on the proximate level. In addition, risk perception in the male is linked with the fact that investment might be in vain. Another male may compete and then investment may be lost. This could at least explain why males make the first overt move in the courtship game. Thus male risk perception depends on the quality of the partner and is modified by possible competition.

A female's risk perception should be governed by ultimate considerations, because of their high investment in the offspring. Females actually appear to follow the hypothetical rules for mate selection, if high status males are better protectors and providers. In contrast to the males, females show higher aspiration in mate choice: they choose actively, elicit approaches, and control the further course of courtship. In addition, female risk-perception is modified by possible behavioural tendencies of the partner, like philandering.

Females change the male's risk-perception through solicitation and eliciting attention. At the same time, females induce male–male competition. High attractiveness of a female seems to imply that she has wider possibilities for mate choice. The male thus supposes that he has to face a great amount of competition with other males and that higher investment is necessary. This makes it likely that the decision for an actual approach is determined by risk-perception. The risk itself is generated on the basis of the attributes of the target person and the amount of solicitation performed.

We could also imagine that cognitive factors are responsible for assortative

161

mating. What has been completely neglected in research to date is the self-assessment of attractiveness. This may augment or reduce risk in a given case. If an individual searches for a partner dependent on the maximum risk he/she can deal with, his/her own qualities will influence risk in such a way that assortative mating occurs automatically. He/she only attains partners according to his/her risk reducing potential.

Human mate choice thus shows two main decision lines. Mates seem to be chosen according to ultimate necessities; an eventual approach appears to be dictated by the risk of rejection. This forces choice to an optimum level.

The actual process of decision-making is another open question. There is evidence, however, that a person is able to 'filter out' at the various stages of courtship those persons who would be unsuitable intimates, by attending to a sequence of differently-based information about a partner (Duck, 1977). This process is complicated by the possibilities of weighing or ranking traits. Males could select a ranking method, because they rely more or less on one trait. Amongst females, however, we find there are numerous traits and thus combinations which could result in the same outcome of fitness for their offspring. Here, however, data on variability in choice, weighing and ranking of factors are lacking.

It is clear that the central problem of this discussion is the weighing of risk-determining factors. Weighing is the tactic an individual will probably choose if there is competition. The possibility of selecting between different traits which might produce the same outcome for fitness could reduce female competition. How far such processes play a role in mate-choice has not yet been touched upon by research. We could postulate that the single concept of attractiveness is divisible into the factors producing it. This would allow us to take into consideration and compare the different factors of attractiveness and reduction of competition amongst males and females.

Furthermore, risk can be modified by a series of factors, for instance individual goal conception. Searching for a partner for fun times or a marriage partner can change risk dramatically. Other factors are individual motivation or self-esteem. Males with high self-esteem (whatever their own physical attractiveness may be) will tend to seek out females of high physical attractiveness, as shown by Stroebe (1977). Thus we should expect flexibility in risk-perception, taking into account historical and societal changes. In other words, changes in the cultural definition of risk of mate-choice itself.

Here, it has been shown that courtship is structured by a set of general rules from which there is little deviation. Cultural and historical influences are mostly neglected in research. They must be taken into account, however, in order to demonstrate the basic components of courtship, which seem to be culturally invariant.

9.6.2 Internal and external organization of strategies

Diminishing social distance and creating affiliation are the main objects of courtship. Risk can be met in manifold ways. We should expect that individuals act according to the risk they perceive. In what way the establishment of power does play a role, however, we do not know. We should expect that for higher status males the risk should be lower.

Symonds (1972) and Cook (1981) both describe the same principle for the negotiation process in courtship. This principle is the variation of verbal and non-verbal directness with perceived chance of success. Here we find explicit sex differences. Females who have an overall higher investment should encounter an overall higher risk. Females are more indirect and so employ non-committal non-verbal invitation, in this way exerting control. Interpretation of the message is left to the receiver and thus response is not obligatory. The overt verbal efforts of males vary according to the risk of rejection. In verbalization, males tend to become indirect under high risk conditions. It is at this point that we find the meshing of verbal and non-verbal signals. Non-verbal behaviour is highly prominent in courtship, because it does not put others in a position of being obliged. So females appear 'animated', but at the same time the situation is one of high ambivalence. For example, the sign-value of a body movement can be doubled by three processes: first, a non-verbal behaviour can be framed by a typical movement configuration. Usually we find a fast start of the movement up to its peak. At the peak a detectable pause occurs. Finally the movement is accomplished by a slow move back to the starting point. Another way is by producing the movement in rhythmical bouts (e.g. nodding). Thirdly, it is possible to change the movement's background, for contrast.

Moreover, we have to admit that the data on verbal behaviour, probably the vain social tool in courtship, remain little more than fragmentary. Thus research on this neglected point is a major issue for further research.

It is not clear if consorts calculate risk out of social distance, the goal and the distribution of power in the dyad. But it is clear that risk assessment occurs. It is certainly a cognitive tool for cost-benefit calculations. Risk seems to be an anticipation of possible benefits and necessary investment, determined by a complex, and probably individually variable, summing up of factors. On the other hand, risk also seems to determine the decisions and the necessary tactics.

Thus, the point to be investigated is: who structures what? At first we find a clear but subtle initiation role on the female's side. This finding is in contrast to the general view of gender roles. The overt initiator is still the male, possibly pressed by male–male competition. Female initiation only lowers risk in the perception of the male. Owing to intersexual conflict, one would expect that the sex with the higher cost should structure the interaction to a higher degree. Indeed, females try to force the contact in the direction of a long-term relationship and press for commitment.

Further evidence for intersexual conflict can be seen in the high degree of ambivalence in the female's non-verbal behaviour. This could well be a means of testing the male's willingness for investment, i.e. coyness. We do not find an active or passive role: both partners structure courtship. But they do it with different means at different times.

Up to this point, we assumed the existence of at least a minimal mutual interest. But what if one of the partners recognizes that the other does not comply with his/her wishes? Here we would hypothesize that verbal behaviour would become more indirect or vague and that the favoured tactics would show either no solicitation behaviour or non-verbal signs of rejection. Another way for the partners would be to impose their wishes aggressively. Surprisingly, Kirkendall (1961) reports that it is the females who often start aggressive acts if the male does not meet with their demands. On the other hand, we find an interesting dichotomy: males use power for attaining, females for avoiding sexual intercourse. This again underlines the fact that the sexes assess the sexual act differentially, with the greater risk on the female side.

All authors studying human courtship emphasize the 'invariance' of courtship. 'Variance' on the other hand is not described. Invariance could well be a result of the interaction of two adaptive systems. We find that large parts of the decisions made and of the repertoire used seem to be influenced or acted out according to biological necessities. On the other hand, we find that actual behaviour and the course of the episodes are structured by another adaptive system. This system has its roots in cognitive restraints on interactions and goal-reaching attempts. Adding to the complexity of the system is the fact that risk has to be met by behaviour tactics with a predetermined function. We can speculate that the higher the risk becomes, the higher the degree of invariance of the courtship episodes.

What is still missing in research is observation of the tactics males and females could use for lowering risk. One of these tactics is certainly self-presentation. Self-presentation with its components could play the main role in human courtship. In self-presentation we should expect sex differences according to the biological hypotheses stated above. Moreover, self-presentation should occur in relation to risk perception. Out of the existing fragmentary work we could define at least five main themes:

(a) The demonstration of qualities in the developing relationship. Status differences and social distance can be manipulated. They can be maximized or minimized with respect to the target person.
(b) The demonstration of number and quality of social relationships outside the dyad (e.g. showing the number of friends).

(c) The demonstration of physical and psychic abilities and knowledge (e.g. accentuation of physical appearance, or showing one's driving ability).

(d) The demonstration of the available object world (the things I have or give away).

(e) The demonstration of the value of intended or executed acts (I did it this way, or we could do this that way).

These dimensions become self-presentation through qualitative changes: over-accentuation and under-accentuation. The first possibility is present in display, the second in ingratiation.

If we introduce self-presentation into the biological hypothesis we expect marked sex differences. A female's self-presentation should mark her attractiveness to males and her offspring-raising potential. Male self-presentation should be adapted to the female's wish for male investment. Resulting from intra-male competition, self-presentation should be performed with higher frequencies among males. Self-presentation tries to lower risk in the perception of the partner at any stage of courtship, and thus should form a central point in research on courtship.

Although we find many hints, it is not yet clear how ingratiation and display are produced in behaviour with respect to risk-perception. Finding out more about this is the aim of a current research project. On the basis of the above mentioned theoretical considerations, we can state the following hypotheses:

1. An approach depends on a comparison of necessary costs, risk and the possible benefits. It might also be controlled by a time limit. The decision for an approach could depend on comparisons of risk and available means. The attractiveness of the actor may play an important role in risk-reduction.

2. The amount of risk controls the amount of solicitation in the female and determines the directness of the male's first move.

3. Tactics in courtship follow the 'good move principle': males and females act according to a perceived risk. If necessary, they create detours which find their expression in verbalization in direct terms.

4. The content of self-presentation is dictated by the ultimate necessities in courtship and is thus different for both sexes.

5. The quality of self-presentation varies with risk-perception, from direct presentation in low-risk situations to indirectness in high-risk situations.

6. The intensity of self-presentation is controlled by female solicitation.

7. Females and males pursue sex-specific information-gathering strategies.

According to the research to date, courtship thus might really be the bastion for gender role-performance. Most of the facts coincide with the hypotheses introduced above. At this point, however, care must be taken since the data base is not yet broad enough. The review presented here relies on a handful of results gathered mostly by interviewing American and Northern European males and females. As already mentioned, interviews might well only continue the stereotypes found that conceptualize shared cultural knowledge.

ACKNOWLEDGEMENTS

This work is part of the research project 'Male courtship: behavioural strategies of self-presentation' granted by the Deutsche Forschungsgemeinschaft (EI 24/11–1).

REFERENCES

Barash, D.P. (1977) *Sociobiology and Behavior* Elsevier, New York.

Bateman, A.J. (1948) Sexual selection, natural selection and quality of advertisement. *Biol. J. Linnean Soc.* **17**, 375–93.

Berk, B. (1977) Face saving at a single's dance. *Soc. Probl.*, **24**, 5, 530–44.

Berscheid, E & Walster, E. (1974) Physical attractiveness. *Adv. Exp. Soc. Psychol.*, **10**, 27–9.

Broverman, I.K., Vogel, S.R., Broverman, M.D., Clarkson, E.E. and Rosenkrantz, P.S. (1972) Sexrole stereotypes: a current appraisal, *J. Soc. Issues*, **28**, 45–51.

Brown, P. and Levinson, S. (1978) Universals in language usage: politeness phenomena. In: *Questions and Politeness* (ed. E. Goody,) Cambridge University Press, Cambridge.

Cant, J.G.H. (1981) Hypotheses for the evolution of human breasts and buttocks. *Am. Natur.*, **117**, 199–204.

Cary, M.S. (1976) Talk? Negotiation for the Initiation of Conversation between the Unacquainted. PhD Dissertation, University of Pennsylvania.

Charlesworth, W.R. (1978) Understanding the other half of intelligence. *Social Science Information*, **17, 5**, 231–77.

Chisholm, J.S. (1976) On the evolution of rules. In: *The Social Structure of Attention.* M.R.A. Chance and R.R. Larsen, John Wiley, London, pp. 235–52.

Clarke, A.C. (1952) An examination of the operation of residential propinquity as a factor in mate selection. *Am. Soc. Rev.*, **32**, 17–22.

Cook, M. (1981) Social skill and human sexual attraction. In: *The Bases of Human Sexual Attraction* (ed. M. Cook), Academic Press, New York, pp. 145–77.

Coombs, R.H. and Kenkel, W.F. (1966) Sex differences in dating aspirations and satisfaction with computer selected partners. *J. Marriage and the Family*, **28**, 62–6.

Cranach, M., Kalbermatten, U., Indermühle, K. and Gugler, B. (1980) Zielgerichtetes Handeln, Huber, Bern.

Darwin, C. (1871) *The Descent of Man and Selection in Relation to Sex.* John Murray, London.

Davis, J.D. (1978) When boy meets girl: sex roles and the negotiation of intimacy in an acquaintance exercise. *J. Personality Soc. Psychol.*, **36**, 684–92.

References

Dawkins, R. and Carlisle, T.R. (1976) Parental investment, mate desertion and a fallacy. *Nature*, **262**, 131–3.

Duck, S.W. (1977) *The Study of Acquaintance.* Saxon House, Farnborough.

Eckland, B.K. (1982) Theories of mate-selection. *Eugenics Quarterly*, **25**, pp. 71–84.

Eibl-Eibesfeldt, I. (1984) *Die Biologie des menschlichen Verhaltens.* Piper, München.

Folkes, V.S. (1982) Forming relationships and the matching hypothesis. *Personality Soc. Psychol. Bull.*, **8, 4**, 631–6.

Ford, C.S. and Beach, F.A. (1952) *Patterns of Sexual Behaviour.* Methuen, London.

Fowler, H.F. (1978) Female choice: an investigation into human breeding strategy. *Paper presented to the Animal Behaviour Society in Seattle, June 1978.*

Frisch, R.E. (1975) Critical weights, a critical body composition, menarche and the maintenance of menstrual cycles. In: *Biosocial interrelations in population adaptation* (ed. E.S. Watts), Mouton, The Hague, pp. 319–52.

Givens, D. (1978) The non-verbal bases of attraction: flirtation, courtship and seduction. *Psychiatry*, **41**, 346–59.

Grammer, K. (1982) Wettbewerb und Kooperation: Strategien des Eingriffs in Konflikte unter Kindern einer Kindergartengrupp e. Dissertation im Fachbereich Biologie an der Ludwig-Maximilians Universität München.

Grammer, K. (1985) Verhaltensforschung am Menschen: überlegungen zu den biologischen Grundlagen des "Umwegverhaltens". In *Mensch und Tier* (ed. M. Svilar), Peter Lang, Bern, New York, pp. 273–318.

Grammer, K. and Shibasaka, H. (1985) Strategies of social manipulation: an ethological view. *Paper presented at the 8th Biennial Meeting of the International Society for Research on Behavioural Development in Tours, France.*

Green, S.K., Buchanan, D.R. and Heuer, S.K. (1984) Winners, losers and choosers. A field investigation of dating initiation. *Personality Soc. Psychol. Bull.*, **10, 4**, 502–11.

Harrison, A.A. and Saeed, L. (1977) Let's make a deal: an analysis of revelations and stipulations in lonely hearts advertisements. *J. Personality Soc. Psychol.*, **35, 4**, 257–64.

Hess, E.H. (1975) *The Tell-tale Eye.* Van Nostrand, New York.

Hinde, R.A. (1984) Why do sexes behave differently in close relationships? *J. Soc. Personal Relationships*, **1**, 471–501.

Homans, G.C. (1961) *Social Behaviour: its Elementary Forms.* Routledge and Kegan, London.

Horvath, T. (1979) Correlates of physical beauty in men and women. *Soc. Behav. Personality*, **7, 2**, 145–51.

Huston, T.L. (1973) Ambiguity of acceptance, social desirability and dating choice. *J. Expl Soc. Psychol*, **9**, pp. 32–42.

Kendon, A. (1975) Some functions of the face in a kissing round. *Semiotics*, **15, 4** 99–334.

Kirkendall, L. (1961) *Premarital Intercourse and Interpersonal Relationship.* Julian Press, New York.

Low, B.S. (1979) Sexual selection and human ornamentation. In: *Evolutionary Biology and Human Social Behaviour. An Anthropologist's Perspective.* (eds. N.A. Chagnon, and W. Irons), Duxbury, North Scituate, Mass., pp. 462–87.

Maynard-Smith, J. (1974) The theory of games and the evolution of animal conflicts. *J. Theor. Biol.*, **47**, 209–21.

McCormick, N.B. (1979) Come-ons and put-offs: unmarried students' strategies of having and avoiding sexual intercourse. *Psychol. Women Q.*, **4**, 194–211.

McCormick, N.B. and Jesser, J.C. (1983) The courtship game. Power in the sexual encounter. In: *Changing Boundaries. Gender Roles and Sexual Behaviour.* (eds. E.R. Algeier and N.B. McCormick), Mayfield, Palo Alto, pp. 64–86.

McDaniel, C.O. (1969) Dating roles and reasons for dating. *J. Marriage and the Family*, **31**, 97–107.

Moore, M.M. (1985) Non-verbal courtship patterns in women: context and consequences. *Ethology Sociobiol.* **6**, 237–47.

Morris, D. (1978) *Der Mensch mit dem wir leben*, Knaur, München.

Murstein, B.I. (1976) *Whom will marry whom? Theories and Research in Marital Choice.* Springer, New York.

Owen, F. (1982) Advertising for a partner – varieties of self-presentation. *Bull. Br. Psychol. Soc.*, **35**, 72.

Parker, G.A. (1983) Mate quality and mating decisions. In: *Mate Choice* (ed. P. Bateson), Cambridge University Press, Cambridge, pp. 141–66.

Perper, T and Fox, V.S. (1980) Flirtation and pickup patterns in bars. *Paper presented at the Eastern Conference on Reproductive Behaviour, New York, June 1980.* Cited in McCormick, N.B. and Jesser, C.J. (1983).

Roebuck, J. and Spray, S.C. (1970) The cocktail lounge: a study of heterosexual relations in a public organization. In: *Studies in Human Sexual Behaviour: the American Scene* (ed. A. Shiloh), Charles Thomas, Springfield, pp. 443–53.

Rosenblatt, P. and Anderson, R. (1981) Human sexuality in a cross-cultural perspective. In: *The Bases of Human Sexual Attraction.* (ed. M. Crook), Academic Press, New York, pp. 215–50.

Rubin, Z. (1973) *Liking and Loving: an Invitation to Social Psychology.* Holt, Rinehart and Winston, New York.

Sack, A.R., Keller, J.F. and Howard, R.D. (1982) Conflict tactics and violence in dating situations. *Int. J. Sociol. Family*, **12, 4**, 89–100.

Scheflen, A. (1965) Quasi-courtship behaviour in psychotherapy. *Psychiatry*, **28**, 245–57.

Sigall, H. and Landy, D. (1973) Radiating beauty: Effects of having an attractive partner on person perception. *J. Personality Soc. Psychol.*, **28**, 218–24.

Skipper, J.K. and Nass, G. (1966) Dating behaviour: a framework for analysis and an illustration. *J. Marriage and the Family*, **28**, 412–20.

Skrzipek, K.H. (1981) Menschliche "Auslösermerkmale" beider Geschlechter. I. Attrapp enwahluntersuchungen der Verhaltensentwicklung. *Homo*, **29**, 75–88.

Skrzipek, K.H. (1982) Menschliche "Auslösermerkmale" beider Geschlechter. II. Attrappenwahluntersuchungen des geschlechtsspezifischen Erkennens bei Kindern und Erwachsenen. *Homo*, **32**, 105–19.

Stroebe, W. (1977) Self-esteem and interpersonal attraction. In: *Theory and Practice in Interpersonal Attraction* (ed. S.W. Duck), Academic Press, London, pp.79–104.

Symonds, L.A. (1972) A vocabulary of sexual enticement. *J. Sex Res*, **8**, 136–9.

Tharp, R.G. (1963) Psychological patterning in marriage. *Psychol. Bull.*, **60**, 97–117.

Thibaut, J.W. and Kelley, K.H. (1959) *The Social Psychology of Groups.* Wiley, New York.

Trivers, R.L. (1972) Parental investment and sexual selection. In: *Sexual Selection and*

References

the Descent of Man 1871–1971 (ed. B. Campbell), Aldine, Chicago, pp. 136–79.

Walster, E., Aronson, V., Abrahams, D. and Rottman, L. (1966) Importance of physical atractiveness in dating behaviour. *J. Psychol. Soc. Psychol.*, **4, 5**, 508–16.

Walster, E., Walster, G., Piliavin, J. and Schmidt, L. (1973) Playing hard to get: understanding an elusive phenomenon. *J. Personality Soc. Psychol.*, **26**, 113–21.

Zetterberg, H.L. (1966) The secret ranking. *J. Marriage and the Family*, **28**, 134–72.

CHAPTER TEN

Reproduction and sex-ratio manipulation through preferential female infanticide among the Eipo, in the Highlands of West New Guinea

Wulf Schiefenhövel

10.1 INTRODUCTION

There can be no doubt that human sexuality and reproduction are classical examples illustrating that many behaviour patterns are firmly biologically based, leaving the respective cultures less latitude to shape actions than in other spheres of life. Sexual desire, eroticism, jealousy, rules regulating celibacy and marriage, bonding with the infant, nursing from a nutritional, emotional, social and intellectual respect, weaning, delaying the possible birth of a sibling, influencing the overall number and sex ratio of the offspring as well as the effects of morbidity and mortality on population dynamics exhibit, particularly when one compares traditional (non-industrial) societies, a host of common characteristics which lend themselves to analysis in the light of evolutionary biology.

Since Darwin (1871), a number of authors, of whom but a few can be mentioned here, have dealt with this topic (Carr-Saunders, 1922; Wynne-Edwards, 1965; Eibl-Eibesfeldt, 1970, 1982; Trivers, 1972; Trivers and Willard, 1973; Daly and Wilson, 1978; Konner and Worthman, 1980; Alexander, 1979; Vogel, 1986). The particular achievement of the sociobiological approach is that it allows the formulation of testable hypotheses. This, amongst other properties, has made the approach attractive for biologists but also for anthropologists and other social scientists.

Human sociobiology suffers from a paucity of field observations on reproduction in, preferably, traditionally living ethnic groups. For many years only the well documented !Kung San data (Howell, 1979) and data from the Yanomamö (Chagnon, 1974, 1979) were available. Of the latter, M. Dickemann writes (1981): 'A more detailed review of the Yanomamö data is . . . critical' because '. . . Chagnon initially believed that he had evidence of sex-ratio manipulation through infanticide, but now contends there is no certain evidence for any Yanomamö infanticide at all'. Recently another case study was published by P. Bugos Jr. and L. McCarthy (1984) who interviewed, through interpreters, Ayoreo informants on reproductive success and infanticide. They detected 54 cases of the latter among 141 births; 38 infants had been male, 17 female, of 7 the gender were no longer known.

During our fieldwork among the hitherto virtually uncontacted Eipo of the Daerah Jayawijaya in the West New Guinean mountains (Figure 10.1) my wife and I were able to conduct a detailed census and to actually observe childbirth and infanticide (Schiefenhövel and Schiefenhövel 1978; Schiefenhövel, 1984, 1986). The observations presented in this chapter may be of particular interest because they stem from an unsegmented traditional society of horticulturists and, with regard to proximate causes (motives) for infanticide, comply with sociobiological theory.

10.2 ETHNOGRAPHIC BACKGROUND

The Eipo, members of the Mek language and culture group (Schiefenhövel, 1976) who live on the northern slopes of the Central Cordillera at an altitude of 1600 to 2100 m, are in many respects typical mountain Papua. In the rather wet climate (approximately 6000 mm rainfall per year) they grow, as

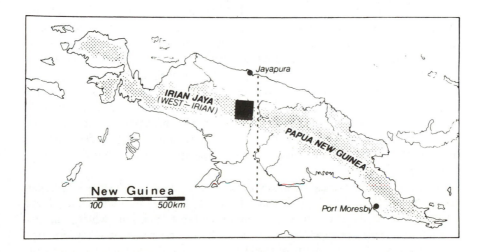

Figure 10.1 New Guinea

171

horticulturists using digging sticks (*kama*), sweet potato (their staple), several green vegetables (the main source of protein), taro and sugar cane. Collecting small animals, together with eggs, larvae, and wild plants is typically done by the women. It contributes a few grams of probably highly valuable protein to their own daily diet and to that of their small children. Pigs, which are almost exclusively used for ceremonial purposes and hunting, are much less important in nutritional terms. It is most likely that the Eipo belong to the old stratum of immigrants to the island, who originated in South-east Asia and, by way of the Indonesian archipelago, arrived on New Guinea approximately 50000 years ago. They have been settled in the favourable mountainous inland for at least 10000 years (Swadling, 1981).

The tools of the Eipo were, apart from very few steel axes and bushknives traded in from distant mission stations, of neolithic type: stone adzes (*ya*) and other instruments made of stone, bone, wood or fibre. Their village had between 30 and 200 inhabitants and consisted of family houses (*dib eik*), men's houses (*yoek eik*) and houses for women (*bary eik*), used during menstruation, childbirth, sickness, and marital tension. The villages were located in such a way that attacks by enemies were difficult or could be easily detected.

'Big men' (*sisinang*) took political and religious initiatives and were followed as long as they acted convincingly; their positions were neither institutionalized nor hereditary. The only professional specialists, who did not necessarily gain material benefit from the service they provided, were (mostly male) healers (*kwetenenang*) and seers (*asing ketenye*). The social structure of this unsegmented, akephalic society was characterized by patrilinear exogamous clans (*yala*), virilocality, potential polygyny, monogamy in most cases, and some very few polyandrous constellations. Male–female social dichotomy was, in Papuan perspective, moderate, i.e. women and men worked and talked together but adhered to their own spheres. Political decisions were usually made by men but women were not necessarily powerless, especially in family affairs. Warfare and intergroup conflict as well as aggression between spouses and within families were not uncommon. Mythic traditions and animistic beliefs pervaded everyday life. Verbal art forms, e.g. rhetoric with coded language and lyrical love songs using rich metaphors, and excellent knowledge of nature (cf. Hiepko and Schiefenhövel, 1987) demonstrate that the Eipo are typical members of our species – curious, intelligent and sensitive.

In 1976 two severe earthquakes took several lives and destroyed villages and, particularly, gardens. Since 1978 a strong influence has been exerted by a North American fundamentalist mission which, in its uncompromising approach and demand to cut the roots of tradition, has already changed a great many of the characteristic cultural traits we were able to study before.

10.3 METHODS

From 1974 to 1976 I spent 17 months in the Eipomek valley and four further months during restudies in 1978 and 1980. Besides carrying out her own dento-anthropological survey, my wife Grete Schiefenhövel participated in the demographic and genealogic study. The interviews with various older and younger male and female informants, mostly of the village of Munggona, were carried out in the Eipo language (a comprehensive dictionary of the hitherto unknown language was published by Heeschen and Schiefenhövel in 1983). The demographic and genealogic investigations were facilitated by the richness of Eipo kinship terminology.

While they are used to counting, the Eipo do not keep numerical track of their age. We therefore assessed that by using, for children up to approximately 12 years, the (New Guinea typical) dentition of deciduous respectively permanent teeth, for persons older than that other dental conditions, especially wear, and additional biographical clues. The respective age ('I was at that time as old as the child X is now') at one of the two short, but vividly remembered visits of the French explorer Pierre Gaisseau in 1959 and 1969 was calculated. This, when checked with other personal data and events (which position in the row of siblings did he/she have – the typical birth-interval being at least three years) gave quite useful hints.

Medical, nutritional, anthropological, human ethological, and historical data were also collected and compared with those of the other scientists in the project.

Owing to the two earthquakes in 1976 and an epidemic after the second, which took further lives, and the mission influence the data are grouped into three periods: Period I: 10/1974 to 5/1976 – this period could be given the label 'neolithic living conditions'; Period II: 6/1976 to 10/1978 (the time of the first demographic restudy) – earthquakes and epidemic; Period III: 11/1978 to 4/1980 – mission influence, which has probably had a special effect on the rate of infanticide. During this period several attempts were made by indigenous mission personnel to reverse the decision of mothers to reject their newborns, and at least one of them was successful.

10.4 RESULTS

Aerial photographs taken by the Dutch in 1945 show (Helmcke, 1983) that since this period the areas with secondary vegetation (gardens and old gardens lying fallow) have neither increased nor decreased, thus indicating that the population in this rather short time span of two decades has remained constant. This inference is backed by our observation, that small patches of primary forest were only rarely cleared to make room for new gardens. Agricultural techniques, using mulch as fertilizer, draining ditches etc., are well developed. It is unlikely that one would, with the materials available, attain a higher yield if a different technology was applied (Plarre, 1978).

Securing sufficient food for the immediate future was a definite preoccupation. This is understandable when one considers that there was, besides the calories stored in the domesticated pigs (which were, however, reserved for payment and ceremonial use) no storage of any kind. The earthquakes which destroyed the majority of the gardens caused severe food shortages. Even in 'normal' times we witnessed one moderate famine in one of the villages which gave rise to a rather impressive relief operation by the inhabitants of the unaffected areas. Our informants told us of severer cases of food shortage in the past. From all this one can conclude that for the period preceding the introduction of steel tools (which of course greatly facilitate the clearing of land) the energy requirements of the population were just below the average carrying capacity of their area.

The population of the six hamlets in the upper Eipomek valley, according to the census of August 1975, is shown in Figure 10.2. The settlement unit Munggona-Kwarelala and the village of Dingerkon represent old villages, whereas the three hamlets forming the settlement unit Malingdam are much younger and developed one or two generations ago from a conglomeration of garden houses. These were situated in a large area of good land which had stayed fallow long enough (approximately 15–20 years) to be reused. Malingdam had, since its foundation, attracted inhabitants who formerly lived in Munggona and the upper Heime valley just across the central range. The latter can be deduced from the fact that quite a number of Hei women were married there, following the typical virilocal residence pattern. The two old and the one young village show marked differences in the sex ratios of young people (the first three age groups). This, as will be discussed later, could well be the effect of conscious decision.

Figure 10.3 shows the sex ratios of the six age groups. The male bias until the age of approximately 20 is quite obvious. Why the tendency towards rebalancing of the number of men and women, which is suggested by the ratios in age group IV and VI, is interrupted in group V is not clear. This seemingly 'odd' value could be due to a historic event unknown to us (long period without war and fewer male deaths? Bad living conditions with more female infanticide?). The sex ratio of those older than 45 follows the worldwise trend: women outlive men. It is therefore likely that the rebalancing of sex-ratios takes place between 20 and 45.

The mortality of infants within the first year of life is shown in Figure 10.4; newborns who were victims of infanticide are excluded. The effect of the earthquakes and the subsequent epidemic (Period II) is quite obvious. The presumably 'normal' infant mortality is probably around 50 per thousand. This is considerably lower than in present day developing countries. Infants and children, breast-fed on demand for two to three years, sometimes longer,

174

	♀							♂					
	Total	I	II	III	IV	V	VI	I	II	III	IV	V	VI
Munggona	127	6	7	3	14	9	15	13	9	8	13	20	10
Kwarelala	66	3	–	1	6	5	11	5	6	4	10	10	5
Dingerkon	60	2	1	1	5	6	5	8	3	6	9	8	6
Imarin	116	11	4	8	11	10	13	9	6	9	13	13	9
Mumyerunde	31	3	–	2	4	4	2	4	1	2	–	7	2
Kabcedama	40	2	1	3	2	2	7	4	4	3	3	5	4
Totals	440	27	13	18	42	36	53	43	29	32	48	63	36

Total population 189 251

Figure 10.2 Population of 6 hamlets in upper Eipomek valley (August 75) by age groups (I=0-5; II=6-11; III=12-19; IV=20-29; V=30-44; VI=45 and over).

175

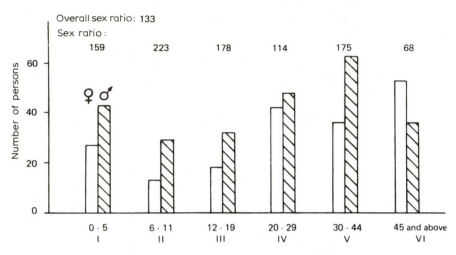

Figure 10.3 Sex ratios (440 persons of 6 hamlets, census 8/75).

	f	m	Total	Livebirths	Mortality
Period I (10/74 — 5/76)	1	—	1	23	1/23 = 43/1000
Period II (6/76 — 10/78)	1	3	4	26	4/26 = 154/1000
(earthquakes, epidemics)					
Period III (11/78 — 4/80)	—	1	1	30	1/30 = 33/1000

Figure 10.4 Mortality of infants within first year of life.

are well cared for and have a great deal of body contact and social stimulation. This archetypic form of childcare is apparently able to compensate for the almost complete lack of medicinal plants.

The upper half of Figure 10.5 contains data on 146 women who were either dead (genealogic data) or past menopause. The sex ratio of those of their children who survived to approximately three years of age is 179 and thus significantly above the expected value. It exceeds that of the women (lower half of Figure 10.5) who lived at the time of the survey and had not yet completed their reproductive lives and also that of the total population (Figure 10.3). This tendency for older women to have relatively fewer girls than younger ones, is also evident from Figure 10.6 (in which only the present generation of women is represented) and constitutes an important finding: women monitor the sex of their offspring, allowing girls to survive more often in the first half of their reproductive lives than in the second. When they become older and have no or few boys they reject female newborns. The small

176

	Total	Female	Male	Unknown sex	
Women dead or past menopause					
Children surviving to appr. 3rd year	398	142	254	2	Total 146
Average number of children	2.6		sex ratio: 179 ($p \leqslant .001$)		

Number of children	0	1	2	3	4	5	6
Number of women	7	23	30	48	23	11	4

	Total	Female	Male	Unknown sex	
Women before menopause					
Children surviving to appr. 3rd year	118	51	65	2	Total 65
Average number of children	1.9		sex ratio: 127 (n.s.)		

Number of children	0	1	2	3	4	5
Number of women	2	29	17	14	2	1

Figure 10.5 Reproductive pattern among the Eipo (data from demographical and genealogical records).

177

	Sex of rejected newborn		
	f	m	*Total*
Women between 20 – 29	4	4	8
Women between 30 – 45	12	3	15
Total	16	7	23

Figure 10.6 Infanticide by younger and older women (observed figures).

number of children, even of mothers beyond menopause, is quite striking. Only four of the mothers had six children who survived to approximately three years of age, the average being 2.6 children per woman. Of the post-menopausal women 78 had 76 daughters who lived to reproductive age themselves; a strong indication of zero population growth, at least for the time period covered.

Figure 10.7 lists all observed cases of birth and infanticide for the three time periods. The drop in the incidence of infanticide in Period III (mission influence) is obvious. Although the sex ratio at birth was 93 and 96 in Periods I–III and I–II respectively it became moderately to considerably male biased after infanticide, an indication that preferential female infanticide is part of the cultural tradition of the Eipo.

We learned that the village community and the husbands and families did not sanction infanticide. There was a general consensus that to accept or reject a newborn was the right of the mother – who could, prior to delivery, have been influenced by others, of course, including the husband. This had clearly happened in one case where the husband had been infuriated by the fact that his wife had become pregnant by her young lover. Yet, notwithstanding the frankness in discussing infanticide, there was a certain ambivalence to be felt. Some of the women, who we knew had rejected their newborns, would either decline to talk about the event or only do so in a circumspect and apologetic manner. When, in the case documented by Grete Schiefenhövel, cries of a very young infant were heard shortly after a mother had carried out infanticide, the other women present immediately calmed the mother down, who had anxiously exclaimed that it had been her baby calling her.

Of those 23 cases of infanticide which happened during the field study, the mothers concerned and/or other women and girls commented on 16 of them, giving the motives listed in Figure 10.8. The last case was that of a very attractive woman of about 20 years, who, by Eipo standards, was at a normal age for the birth of a first child. It seems that she and/or her husband (who was later killed and eaten by the enemy of the neighbouring village) feared the sex taboo during lactation. We never heard of a case where the newborn was rejected because it was male.

178

	Month/year	Total births	f	m	Total inf.	f	m
I	10/74 – 5/76	23	13	10	10	8	2
II	6/76 – 10/78	26	12	14	10	7	3
III	11/78 – 4/80	30	16	14	3	1	2
Totals		79	41	38	23	16	7

Periods I – III	infanticide:	29%
	sex ratio at birth:	93
	sex ratio after infanticide	
	(directly after birth):	124 ($p \le 0.1$;chi^2)
Periods I – II	infanticide rate:	43%
(excluding mission influence)	sex ratio at birth:	96
	sex ratio after infanticide:	190 ($p \le 0.05$ chi^2)

Figure 10.7 Live births and infanticides (observed figures).

Extramarital pregnancy (6 cases);
Specific plan to reject daughter but to accept son (3 cases, in 2 of which the husbands had the same explicit preference for a baby boy);
Malformation of the baby (2 cases of alleged club foot);
Desire to have no more children at all (2 cases);
Infant born too early after the last child (1 case);
Mother considered herself to young to have a child (1 case).

Figure 10.8 Motives (proximate causes) for infanticide in Eipo women (for 15 of 23 infanticides which occurred during fieldwork).

Some weeks after infanticide, ovulation sets in again, and there is no post-partum coitus taboo to be adhered to. In such cases a woman might quickly conceive again. The overall number of her offspring, as can be seen in Figure 10.9, is therefore less reduced than might be expected. For the models, the following conditions were kept constant: first child born at age 20, last at age 44; after each accepted child, lactation induced infertility and post-partum coitus taboo for three years, thus birth of next sibling four years later; no lactation infertility and no coitus taboo after infanticide, thus next birth one year later; first child female; preference for boys. The reproductive strategy sketched in model (a) seems to be very rare among Eipo women. Models (c) and (d) come close to what we have observed with regard to infanticide rate (30%) and sex ratio (190 for Periods I and II). The total number of offspring is, of course, too high in these models. Six surviving children is, as was shown in Figure 10.5, the exception. The models do not make any allowances for shorter reproductive lives (first child later, last child earlier), for miscarriages, disease and other physical and psychological factors delaying conception, etc.

The following conditions were kept constant:

1st child born at age 20, last at age 44; after each accepted child lactation induced infertility and post-partum coitus taboo for three years, thus birth of next child four years later; no lactation-infertility and no coitus taboo after infanticide, thus birth of next child one year later; first child female, preference for boys.

Model (a) 20 y. 1st child, fem.
24 y. 2nd child, male
28 y. 3rd child, fem.   inf.-rate: 0%
32 y. 4th child, male   sex ratio: 75
36 y. 5th child, fem.   7 surv. children, 4 f, 3m
40 y. 6th child, male
44 y. 7th child, fem.

Model (b) 20 y. 1st child, fem.
24 y. 2nd child, male
28 y. 3rd child, fem.;   INF.   inf.-rate: 25%
29 y. 4th child, male   sex ratio: 200
33 y. 5th child, fem.   6 surv. children, 2 f, 4 m
37 y. 6th child, male
41 y. 7th child, fem.;   INF.
42 y. 8th child, male

Model (c) 20 y. 1st child fem.
24 y. 2nd child, male
28 y. 3rd child, fem.;   INF.   inf.-rate: 33%
29 y. 4th child, male   sex ratio: 200
33 y. 5th child, fem.;   INF.   6 surv. children, 2 f, 4 m
34 y. 6th child, male
38 y. 7th child, fem.;   INF.
39 y. 8th child, male
43 y. 9th child, fem.

Model (d) 20 y. 1st child, fem.;   INF.
21 y. 2nd child, male
25 y. 3rd child, fem.
29 y. 4th child, male   inf.-rate: 40%
33 y. 5th child, fem.;   INF.   sex ratio: 200
34 y. 6th child, male;   INF.   6 surv. children, 2 f, 4 m
35 y. 7th child, fem.
39 y. 8th child, male
43 y. 9th child, fem.;   INF.
44 y. 10th child, male

Figure 10.9 Theoretical reproduction models for Eipo women.

The main advantage of the models is, perhaps that they demonstrate that, under the Eipo-typical conditions, infanticide does not serve so much to reduce the overall number of offspring, as to open a very effective possibility of manipulating the secondary sex ratio when one gender is preferred. This appears to be the rationale behind Eipo infanticide.

How, then, do nature and culture deal with the surplus of young men which, as we know from the many fights kindled by extramarital affairs, sexual jealousy and the like, creates tension and social instability? Male sexual frustration was further enhanced by the coitus taboo after birth, to which the wives adhered very strictly. Men were inclined to have lovers during this time, but the shortage of suitable partners (clan exogamy rules are also followed for pre and extramarital intercourse!) rendered this rather difficult. Figure 10.10 summarizes the effects of unbalanced sex ratios on the chances of men finding a wife as a permanent sexual and economic partner. Women are married at about 20, men usually several years later. In contrast to the rather reduced

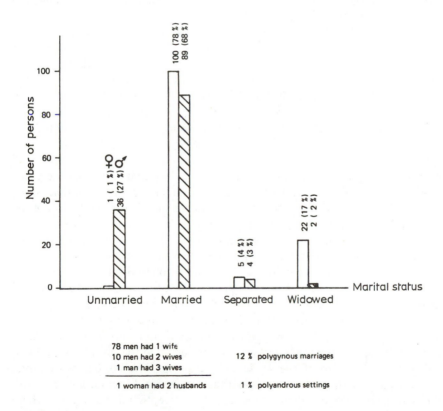

Figure 10.10 Marital status of 128 women and 131 men (total 259), census 8/75.

181

marriage chances for younger men, those for women are excellent: almost every women of under approximately 50 was married; only very few (4%) remained without a husband for a short while after having been widowed, or after divorce, and only the 'really' old widows did not remarry again. This pattern is of course created mainly by the male biased sex ratio of the younger and the female biased one of the older cohorts. One would think that, under the prevailing conditions of woman shortage, polygyny would not have developed or would at least play an insignificant role. Yet, the data show that of 128 women, 20 were spouses of a husband who had two or, in one case, three wives. Had monogamy been the only marital arrangement, 18% more would have been available for marriage to (presumably involuntary) bachelors. Some of these take their fate into their own hands and elope, usually after a love affair, with the second, mostly young and attractive, wife of a polygynist. Frequently the couple marries afterwards. In the long run, male marriage chances are good. They must simply wait, as many of the men do, and be content with a wife older than themselves. Approximately 5% of all men will have to live permanently without a wife.

Which factors influence the process of rebalancing the male surplus? Intergroup warfare (*ise mal*) and the ethnosemantically differentiated intergroup fights (*abala*) are both carried out with bows and (unpoisoned but barbed) arrows. The death toll in warfare was seven per thousand in five years, that of intergroup fighting was, within the Eipomek valley, 3 out of a population of 440 for the same time period, amounting to another seven per thousand. From these figures, one can calculate a kill-rate of three per thousand per year. McArthur (1971) has calculated five per thousand per year for a similar ethnic group in the Eastern half of the island. Figure 10.11(a) lists all observed male and female deaths and indicates in which cases they were due to killing. The rather high losses among the male population at the time of the earthquakes and epidemic were due to accidents and disease and obscure the pattern. In Figure 10.11(b), therefore, Period II is excluded. Males killed then account for 28% of male deaths whereas amongst the women, the percentage killed is 11%. Women are sometimes killed because the community believes them to be witches who can cause disease and death through black magic (*kire*).

The rather low figures in our sample do not lend themselves to statistical analysis. My impression is, however, that boys and men die of disease, accident and armed conflict more often than girls and women. Figure 10.12 attempts to summarize the interplay of biological and cultural factors which jointly influence the number and sex ratio of the offspring not only of individual mothers but also of the whole ethnic group. The coexistence of long breast-feeding with its demonstrable effect on ovarial function together with the post-partum coitus taboo which prevents conception in those cases where ovulatory cycles might have commenced earlier than two years after delivery, is particularly interesting.

	Female deaths	Male deaths	Total deaths
(a) Data from all observed investigation periods (1974-1980)			
n	34	49	83
Killed	3 (9 %)	5 (10%)	8 (10%)
(b) Data from observation periods I and III (figures from times of earthquake and subsequent epidemics excluded)			
n	9	18	27
Killed	1 (11%)	5 (28%)	6 (22%)

Figure 10.11 Percentage of violent death

Biological factors	Outcome	Cultural factors
Late puberty (menarche around 17 years)		
	late 1st pregnancy (usually after 20 years) = shortening of reproductive period	
		– probably low frequency of sexual intercourse (estimate: 1/week)
	lowered chance of conception	
– Lactation up to three years menstruation begins again after appr. 15 mths.)		– post-partum coitus taboo (appr. 2 — 3 years)
	lowered chance for conception	
		– infanticide
	lowered number of children	
		– preferential female infanticide
	reduction of female population	
– Natural death (sickness, accidents, old age)		
	reduction of population (male mortality higher than female)	
		– death in fights and war
	reduction of population (male mortality higher than female)	
		– suicide
	reduction of population (female mortality higher than male)	
		– killing of 'witches'
	reduction of population (female mortality higher than male)	

Figure 10.12 Population control among the Eipo, Highlands of West New Guinea.

The model in Figure 10.13 attempts to explain how male sexual conflict is kept at a more or less tolerable level and overall social harmony is maintained. It seems that the men's houses, so typical for New Guinea highland cultures, with their well structured all-male membership, have a function equivalent to monasteries, where men are given a home and a meaningful place in society.

10.5 DISCUSSION AND CONCLUSION

From our late 20th century Western viewpoint infanticide seems strangely cruel. Yet, infanticide was probably rather common in prehistoric and historic societies and remains so amongst contemporaneous peoples (Scrimshaw, 1984). Despite the educational efforts of the churches (Smith, 1983) feticide-infanticide-pedocide (Dickemann, 1979a) are still part of our own culture as one can see, for example in judicial routine: to kill an infant after birth is usually not regarded as murder but as manslaughter or an even less serious offence (Damme, 1978). The very common practice of induced abortion in our societies and child neglect must be viewed in the same perspective. In Melanesia infanticide was probably much more widespread than anthropologists have been able to discover (Schiefenhövel, 1984). Technical control of conception is rather rare in the traditional societies of the Western Pacific. Oral contraception through herbal medicines is often found (Schiefenhövel, 1970) but its effectiveness doubtful in most cases. Coitus interruptus, inter-femoralis, analis, fellatio and cunnilingus are virtually absent in heterosexual intercourse. The inhabitants know of the causal connections between intercourse and conception but knowledge of the fertile days within the menstrual cycle, as was described for an African group (Jesel, 1986), and thus the conscious directing of coitus toward or away from that optimum, is not described in the Melanesian literature. The Eipo culture is void of all means of controlling conception (except periodic sexual abstinence) through contraception or abortifacient methods. The latter involve, in some ethnic groups, rather effective but also dangerous plant substances. Eipo infanticide can therefore be considered as 'delayed abortion'. In contrast to abortion, it offers some important advantages: there is no medical danger (apart from birth itself) and one has the chance to select newborns according to their physical (and physiological?) status as well as their sex. From a biopsychological aspect, infanticide creates a conflict for the mother, who is, in most cases, the one who carries out the act. Women reject their babies against their basic inclination to become attached to them and care for them (Lorenz, 1943; Bowlby, 1958). In one case (Schiefenhövel and Schiefenhövel, 1978) during her late pregnancy a woman had repeatedly and openly stated that she would not accept another baby girl (because she already had one daughter and no son). After the birth of a healthy female newborn she showed obvious ambivalence. For two hours she deserted her infant, who was still attached to the

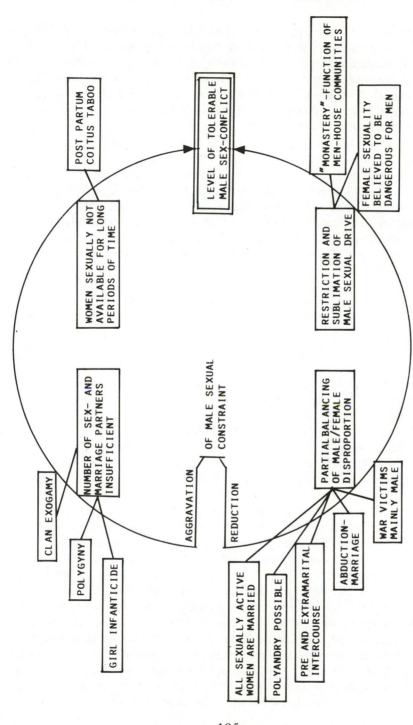

Figure 10.13 Balancing of male sexual conflict.

185

placenta, but then accepted her. She explained later that she had been unable to kill the strong newborn. In the light of attachment theory, the Eipo way of committing infanticide is logical: rejection directly after birth, before there is time for a strong bond to develop, and while the effects of physical and psychological strain connected with birth are still present. Their practice of not cutting the umbilical cord before the placenta appears provides the possibility to react flexibly and rationally. The 10–20 minutes required for the expulsion of the afterbirth give the mother time to judge health, vitality and possibly other characteristics of her child (Figure 10.14).

The motives behind infanticide in certain cases are, amongst the Eipo and also in crosscultural perspective (Daly and Wilson 1984), rather straightforward and fit well with sociobiological theory. Daly and Wilson's list of 'circumstances of alleged infanticide' which was derived from a total sample of 60 societies of which 39 practised infanticide, contains (Figure 10.15) rather similar motives to the ones expressed by the Eipo informants. I am convinced that these motives are the same as those which result in Western mothers and fathers taking refuge in abortion, particularly after amniocentesis and medical analysis to detect trisomia and other fetal defects has given positive results. The Daly–Wilson list does not contain 'preference for a specific gender of the newborn' as a motive. This may, however, also be an important factor. For some segmented cultures, Dickemann (1979a) found indications that preferred female infanticide, e.g. in high Indian castes, follows sociobiological hypotheses: in elite classes, where male offspring may have particularly good reproductive success, newborn girls are rejected, whereas in the lower caste this is absent. It remains unexplained why in these latter classes there is no preferential *male* infanticide.

The Eipo society is pronouncedly unsegmented and nonetheless shows a marked bias for male offspring. One could argue, in an effort to save the inclusive-fitness line of argument, that chances for Eipo daughters might be better when they are supported by a number of brothers and do not have to compete with other sisters for this brotherly help. Our observations for the Eipo do not fit this picture, whereas sisters are helped by their brothers amongst matrilinear Trobrianders (Bell-Krannhals and Schiefenhövel, 1986). The analysis of our demographic data does not offer, as can be seen from Figure 10.16, an explanation within this theoretical framework. When the number of brothers (upper half of table) and sisters (lower half) is plotted against the number of surviving children, the evolving pattern is contrary to the sociobiological assumption: neither brothers nor sisters facilitate own reproductive success.

Amongst the Eipo, the individual decision of the mother to either accept or reject a newborn somehow ties in with culturally transmitted norms. It leads to a reduction in the number of women, who would otherwise grow up and become the mothers of new children, and thus has a strong effect on the reduction of overall population growth. The ecological balance, as one can

Figure 10.14 (a) An Eipo woman, approx. 30 years old, has just given birth to her third child, a girl, whom she rejected. Her firstborn was a girl, the second a boy. (b) In the usual infanticide method child, fernleaves and placenta are firmly tied together as soon as the latter has appeared. A simple bark fibre (visible under the left arm) is used. The bundle was then thrown into an impenetrable thicket of *Saccharum spontaneum* (photographs: *Grete Schiefenhövel*).

187

Society	Inappropriate paternity			Inadequate parental resource circumstance						
	Adulterous conception	*Nontribal sire*	*Sired by mother's first husband*	*Poor infant quality – deformed or very ill*	*Twins*	*Birth too soon or too many*	*No male support*	*Mother dead*	*Mother unwed*	*Economic hardship*
Africa										
Dogon	X			X				X		
Twi	X			X		X			X	
Tiv							X			
Baganda				X						
Masai				X						
Pygmies					X					
Azande									X	
Bemba				X						
Lozi					X					
Asia										
Central Thai				X						
Andaman		X		X	X					
Europe										
Serbs	X								X	
Lapps										
Middle East										
Somali				X					X	
North America										
Tlingit	X				X	X				
Copper Eskimo				X	X	X			X	
Blackfoot				X						
Ojibwa	X						X		X	

188

	15	3	2	21	14	11	6	6	14	3
Iroquois								X		
Klamath	X			X				X	X	
Tarahumara	X			X				X	X	X
Oceania										
Iban	X			X			X	X	X	
Toradja	X			X	X		X	X	X	
Aranda	X	X		X	X		X		X	X
Trobriands										
Lau										
Truk	X			X	X	X			X	
Tikopia	X		X	X	X		X	X	X	X
Russia										
Yakut										
Chukchee				X			X	X		X
South America										
Cuna	X	X		X	X	X	X	X	X	
Cagaba										
Aymara	X			X	X	X				
Ona					X	X				
Mataco	X			X	X		X	X	X	
Guarani				X	X					
Bororo				X	X					
Yanomamo	X		X	X	X					
Tucano	X			X	X				X	
Number of societies	15	3	2	21	14	11	6	6	14	3

(a) Circumstances in which infanticide allegedly occurs in 39 out of 60 societies in a representative sample drawn from the Human Relations Area File. Listed are 95 infanticidal circumstances that make clear reproductive strategic sense for the parents: other miscellaneous rationales are discussed in the text. (A bibliography of ethnographic materials from which this table was compiled is available on request from the authors.)

Figure 10.15 Circumstances of alleged infanticide in society [a] (from Daly and Wilson, 1984).

189

Number of brothers	0	1	2	3	4	Total
Women (past reproductive stage)	4	11	12	8	3	38
Children surviving past childhood	14	31	28	19	6	98
Average number of surviving children	3.5	2.8	2.3	2.4	2.0	overall average 2.6

Number of sisters	0	1	2	3	4	Total
Women (past reproductive stage)	22	11	5	—	—	38
Children surviving past childhood	63	25	10	—	—	98
Average number of surviving children	2.9	2.3	2.0	—	—	overall average 2.6

Figure 10.16 Reproductive success in relation to number of brothers and number of sisters (data mostly from genealogical records).

deduct from aerial photography, historic findings, and observations of actual behaviour, seems to be maintained in this way. This resource-availability viewpoint has probably not been the only motive for population control. Social stability and harmony, fostered by no or little population growth might have been equally important for early (and modern?) Man.

The impression that Eipo infanticide serves supraindividual benefits is convergent with the findings of Bugos and McCarthy (1984). In their study on the Bolivian and Paraguayan Ayoreo Indians, where there are more male than female newborns killed, they state: 'It is difficult to reject on strictly empirical grounds the hypothesis that Ayoreo marriage formation practices (which are, according to them, influenced by infanticide) were group adaptations to regulate population growth.' Sociobiological theory has to explain how, on a crosscultural basis, particular reproductive strategies in general and sex-ratio manipulation in particular, can be due to a number of factors. We might, despite some in-depth anthropological research among traditionally living ethnic groups and very good statistical data for modern societies, simply not know enough about the different child rearing practices and their consequences for demography. Evolutionary biology, in its sociobiological costume, might not yet have developed sufficiently detailed and general theories to cover human reproduction. Some of its assumptions may not have the same stringency for human breeding groups as it has for other mammals. It is also conceivable, as Eibl-Eibesfeldt suggests (1982), that the unit of selection may, in some cases and for some traits, be the whole group and not just the kin.

The various attempts of biologists to shed light on the principles of human reproduction, which are to be carried out in partnership with anthropologists and demographers (Caldwell *et al.*,1987), are not only completely justified but are also a step forward from the sometimes rather cultural-relativistic views towards a more unifying theory.

ACKNOWLEDGEMENTS

I am indebted to my wife Grete Schiefenhövel for her co-operation in the field and a number of other persons and institutions, among them G. Koch, K. Helfrich, S. Gunawan, the Deutsche Forschungsgemeinschaft, LIPI and, above all, the Eipo themselves. For eliminating Germanisms and other mistakes in the manuscript I thank Anne Rasa.

REFERENCES

Alexander, R.D. (1980) *Darwinism and Human Affairs*, Pitman, London.
Bell-Krannhals, I.N. and Schiefenhövel, W. (1986) Repu et de bonne reputation – Système de partage de yam aux îles de Trobriand, Nouvelle-Guinée. *Ecologie-Ethologie Humaines*, **5**. 128–40.

Bowlby, J. (1958) The nature of the child's tie to his mother. *International Journal of Psychoanalysis*, **39**, 350–73.

Bugos, P.E. and McCarthy, L. (1984) Ayoreo infanticide: A case study. In: *Infanticide*(eds G. Hausfater and S. Blaffer-Hrdy) Aldine, New York, pp. 503–20.

Caldwell, J., Caldwell, P. and Caldwell, B. (1987) The mutual reinforcement of speculation and research. *Current Anthropology*, **28** (1), 25–34.

Carr-Saunders, A.M. (1922) *The Population Problem: A Study in Human Evolution.* Clarendon, Oxford.

Chagnon, N.A. (1974) *Studying the Yanomamö.* Holt, Rinehart and Winston, New York.

Chagnon, N.A. (1979) Sex-ratio variation among the Yanomamö Indians. In: *Evolutionary Biology and Human Social Behavior: An Anthropological Perspective.* (eds N.A. Chagnon and W. Irons), Duxbury, North Scituate, Mass., pp. 290–320.

Daly, M. and Wilson, M. (1978) *Sex, Evolution and Behaviour.* Willard Grant, Boston.

Daly, M. and Wilson, M. (1984) A sociobiological analysis of human infanticide. In: *Infanticide* (eds. G. Hausfater and S. Blaffer-Hrdy), Aldine, New York, pp. 439–62.

Damme, C. (1978) Infanticide: The worth of an infant under law. *Medical History*, **22**, 1–24.

Darwin, C.G. (1871) *The Descent of Man and Selection in Relation to Sex.* Appleton, New York.

Dickemann, M. (1979a) Female infanticide, reproductive strategies and social stratification: A preliminary model. In: *Evolutionary Biology and Human Social Behavior: An Anthropological Perspective* (eds N.A. Chagnon and W. Irons), Duxbury, North Scituate, Mass., pp. 321–67.

Dickemann, M. (1979b) The ecology of mating systems in hypergynous dowry societies. *Social Science Information*, **18**, 163–95.

Dickemann, M. (1981) Maternal infanticide and dowry competition: A biocultural analysis of purdah. In: *Natural Selection and Social Behaviour: Recent Research and New Theory*(eds R.D. Alexander and D.N. Tinkle), Chiron, New York, pp. 417–38.

Eibl-Eibesfeldt, I. (1970) *Liebe und Hass. Zur Naturgeschichte elementarer Verhaltensweisen.* Piper, München.

Eibl-Eibesfeldt, I. (1982) Warfare, man's indoctrinability and group selection. *Zeitschrift für Tierpsychologie*, **60**, 177–98.

Heeschen, V. and Schiefenhövel, W. (1983) *Wörterbuch der Eipo Sprache: Eipo-Deutsch–Englisch.* Reimer Berlin.

Helmcke, D. (1983) *Die Trimetrogon-Luftbilder der USAF von 1945 – Die ältesten Dokumente über das Eipomek-Tal und seine Umgebung (West-Neuguinea, Indonesien).* Reimer, Berlin.

Hiepko, P. and Schiefenhövel, W. (1987) *Mensch und Pflanze. Ergebnisse ethnotaxonomischer und ethnobotanischer Untersuchungen bei den Eipo.* Reimer, Berlin.

Howell, N. (1979) *Demography of the Dobe !kung.* Academic Press, New York.

Jesel, R. (1986) *Kulturelle und biologische Aspekte generativen Verhaltens in ostafrikanischen Gesellschaften.* Breitenbach, Saarbrücken.

Konner, M. and Worthman, C. (1980) Nursing frequency, gonadal function and birth spacing among !kung hunter-gatherers. *Science*, **207**, 788–91.

Lorenz, K, (1943) Die angeborenen Formen möglicher Erfahrung. *Zeitschrift für Tierpsychologie*, **5**, 235–409.

McArthur, M. (1971) Men and spirits in the Kunimaipa valley. In: *Anthropology in*

References

Oceania (eds R.L. Hiatt and C. Jayawardena), Angus and Robertson, Sydney.

Plarre, W. (1978) Forschungsprojekt Biologie der Kulturpflanzen. *Steinzeit heute,* **26** (Berlin, Museen preussischer Kulturbesitz).

Schiefenhövel, W. (1970) Die Anwendung von Heilpflanzen und die traditionelle Geburtenkontrolle bei Eingeborenen Neuguineas. *Sitzungsberichte der Physikalisch-medizinischen Sozietät zu Erlangen,* **83, 84,** 114–33 and 179–82.

Schiefenhövel, W. (1976) Die Eipo-Leute des Berglands von Indonesisch-Neuguinea. *Homo,* **26** 263–71

Schiefenhövel, W. (1984) Preferential female infanticide and other mechanisms regulating population size among the Eipo. In: *Population and Biology* (ed. N. Keyfitz), Ordina, Liège, pp. 169–92.

Schiefenhövel, W. (1986) Populationsdynamische Homöostase bei den Eipo in West-Neuguinea. In: *Regulation, Manipulation und Explosion der Bevölkerungsdichte* (ed. O. Kraus), Vandenhoeck and Ruprecht, Göttingen, pp. 53–72.

Schiefenhövel, G. and Schiefenhövel, W. (1978) Eipo, Irian Jaya (West-Neuguinea) Vorgänge bei der Geburt eines mädchens und Änderung der Infantizid-Absicht. *Homo,* **29** 121–38.

Scrimshaw, S. (1984) Infanticide in human populations: societal and individual concerns. In: *Infanticide* (eds. G. Hausfater and S. Blaffer-Hrdy), Aldine, New York, pp. 439–62.

Smith, P.K. (1983) Biological, psychological, and historical aspects of reproduction and child care. In: *Animal Models of Human Behavior.* (ed. G.C. Davey), John Wiley and Sons, Chichester, pp. 159–77.

Swaddling, P. (1981) *Papua New Guinea's Prehistory.* Natural Museum and Art Gallery, Boroko, Papua New Guinea.

Trivers, R.L. (1972) Parental investment and sexual selection. In: *Sexual Selection and the Descent of Man, 1871–1971.* (ed. B. Campbell), Aldine, Chicago, pp. 136–79.

Trivers, R.L. and Willard, D.E. (1973) Natural selection and parental ability to vary the sex-ratio of offspring. *Science,* **179,** 90–92.

Vogel, C. (1986) Populationsdichte-Regulation und individuelle Reproduktionsstrategien in evolutionsbiologischer Sicht. In: *Regulation, Manipulation und Explosion der Bevölkerungsdichte.* (ed. O. Kraus), Vandenhoeck and Ruprecht, Göttingen, pp. 11–30.

Wynne-Edwards, V.C. (1965) Selfregulating mechanisms in populations of animals. *Science,* **147,** 1543–8.

CHAPTER ELEVEN

Women's reproduction and longevity in a premodern population (Ostfriesland, Germany, 18th century)

Eckart Voland and Claudia Engel

11.1 INTRODUCTION

In terms of reproductive strategy, there is an antagonistic relation between fertility and longevity, because they are based – at least partly – on the same limited resources. As most multicellular organisms are not designed to live forever, they are inevitably confronted with the problem of allocating their resources optimally for reproductive and/or maintenance purposes. Natural selection assesses the various allocation solutions according to their effects of fitness, thus rewarding a certain trade-off between fertility and survival by fixing it genetically in the population.

There are valid indications of the accuracy of this hypothesis from inter-species comparisons. Ricklefs (1977), for example, was able to demonstrate how an increase in fertility correlates with a decrease in survivorship of the breeders among birds. Also, similar confirmations are to be found in intra-species analyses (Bryant, 1979, for house martins; Fedigan *et al.*, 1986 for Japanese macaques).

Daly and Wilson (1983) correctly point out that contradictory results (i.e. positive correlations between fertility and survivorship or longevity) do not automatically lead to a falsification of the hypothesis, because organisms are known to adjust their reproductive efforts to current and individual resource situations. Reproductive effort and survivorship can be influenced equally and to the same degree by the prevailing ecological conditions and, for that reason alone, a positive correlation between the two is likely. The positive correlation between clutch size and parental survival among magpies, for example, can be explained by the mutual dependence of both variables on the quality of the territory (Högstedt, 1981).

The 'cost of reproduction' hypothesis (Hirshfield and Tinkle, 1975; Williams, 1966; Tinkle, 1969) is derived from the intraindividual competition of reproductive effort and somatic effort for limited resources. It is assumed that costs arise for the individual through the use of resources for a reproductive event and that these costs decrease the individual's capacity for maintenance and future reproduction.

Calow (1979) provides numerous examples of the physiological basis of the accruing costs. Among isopods (as well as among many other plant and animal organisms) the 'detour' of metabolic materials and energy to the production of gametes leads to the retarded growth of the parent individuals and, therefore, to a decreased prospect of reproductive output (Lawlor, 1976). Among red deer, milk kinds have significantly less fat reserves and a correspondingly lower probability of surviving the following winter than comparable yeld hinds (Clutton-Brock *et al.*, 1983), and gravid lizards run a higher risk of becoming the victim of birds of prey and other predators, due to their reduced mobility (Shine, 1980).

This, of course, has consequences for the fitness of an organism, so that the phenotypic costs of reproduction in terms of energy, biological substance and risks to life can also be measured in the evolutionary currency of residual reproductive value (Fisher value) (Gadgill and Bossert, 1970). Current reproduction, according to the theory, reduces the prospect of future reproduction.

As plausible and suggestive as the 'costs of reproduction' hypothesis may be, critical literature reviews by Shine (1980), Galef (1983) and Bell (1984a) illustrate – maybe with the exception of reptiles – how little empirical evidence there really is for this theory. The results of the numerous field and laboratory studies, as well as of controlled experiments available to date definitely do not all point towards an unequivocal confirmation. Even Bell's (1984a,b) systematic studies on asexual freshwater invertebrates contradict the existence of reproductive costs; however, under standardized (i.e. constant and comparable) conditions, Bell's data illustrate almost without exception a positive correlation between present reproduction and future reproduction or survival, respectively.

As is generally known, the residual reproductive value is determined by two components, namely by fecundity and the probability of survival. For this reason, the reproductive costs can be divided into fecundity costs and somatic costs for analytic purposes. This differentiation is especially important in connection with this study, because we are trying to determine the possible phenotypic costs of reproduction for women. Human females seem to be the only organisms for whom survival and fecundity do not systematically correlate with each other in the higher age classes, due to the achievement of menopause. This is why possible reproductive costs can be expressed in phenotypically various and mutually independent currencies.

In this study (Voland and Engel, 1986), we are interested in the correlation between fertility and longevity, and first of all, in the question of whether or not there are indications of somatic (survival) costs in a human population which can be derived from women's reproductive histories. In empirical surveys on the determinants of lifespan, the reproductive history of women as an independent variable was only taken into account in a few cases, such as in the papers Arvay and Takàcs (1966) and Wyshak (1978). The older studies were summarized by Lawrence (1941). All of these studies had unexpected results; they more or less clearly established a positive correlation between female fertility and longevity. It must be explicitly mentioned that we are not referring to the trivial fact that women must live longer in order to be able to bear many children. Instead, the correlation exists for women of post-menopausal lifespans. It is only for extremely fertile women with ten to more than 20 children that there is any evidence for a life-shortening effect from reproductive activity. These women experience a below-average lifespan (Powys, 1905).

No comprehensive and satisfactory interpretation of these findings was offered in any of the publications cited above. That is why we first wish to probe the relationship between fertility and longevity, in order to explore subsequently the significance of this relationship from an evolutionary point of view.

11.2 MATERIAL AND METHODS

Demographic data from a historical population of Ostfriesland (Germany) were analysed The initial material consisted of the completely known life histories of 811 women, who were all born between 1700 and 1750, married and survived at least until their 47th birthdays. A life history is considered to be completely known if the date of birth or baptism, the dates of marriages, the date of death or of the funeral, and the number of birth dates of the children out of all marriages are known exactly at least to the year. In those cases, in which only the year but not the day is exactly known, the middle of the year has been arbitrarily set as the date of the event, so that the maximum error does not exceed half a year for the respective dates.

We obtained the data relevant to our enquiry from the *Ostfrieslands Orts-sippenbücher* published by the Ostfriesische Landschaft (1966–1984) for the seven parishes of Hesel, Loga, Middels, Ochtelbur, Reepsholt, Werdum and Westerende. The data were then transferred to a SPSS system file and evaluated with the help of various statistical routines available in the program package. Other statistical analyses were performed according to Sachs (1978). More details on the methods used are to be found in Voland and Engel (1986).

11.3 RESULTS

11.3.1 Longevity and reproductive effort

For our sample the relation between the further life expectancy of 47 year-old women (e_x) and their reproductive effort is presented in Figure 11.1. It considers (i) the number of all the children who were born (including still-births), (ii) the number of those born alive and (iii) the total number of deliveries (twin births counted singly).

The curves show some pecularities that need to be interpreted; for example the finding, which contradicts that of Arvay and Takàcs (1966), that married but barren women live longer than their fertile counterparts. Furthermore, the below-average life expectancy of those women who gave birth to two children is remarkable. Apart from the latter group, there does not appear to be any obvious correlation between fertility and the life expectancy of post-menopausal women. If the children survive, there is, however, a 'monotonous trend' in the increase in the lifespan of the women with the number of surviv-

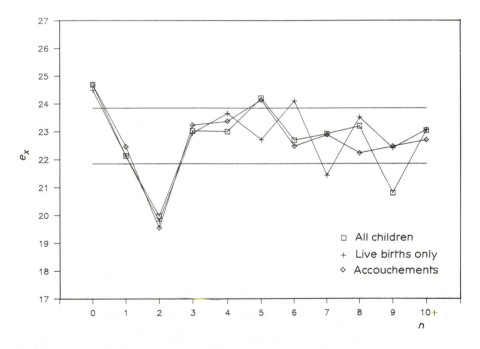

Figure 11.1 Future life expectancy of 47-year-old women (e_x) by number of accouchements, all children (including stillbirths) and live births. The horizontal bars mark the 99% confidence interval for the population mean ($x = 22.85$).

ing children ($r_s = 0.91$, $p < 0.005$, Figure 11.2). (Surviving children are those who attain the age of at least 15.)

The presumed life-shortening, physiological effects of reproductive activity can only be supposed for our sample, insofar as reproductive behaviour basically decreases life expectancy but only to a very slight extent. The amount of reproduction is obviously unimportant, i.e. the survival costs take effect with the first pregnancy, and they are by no means less than those for repeated gravidity. The renunciation of genetic reproduction enables barren women to make an above-average somatic effort, which is expressed by a slightly higher life expectancy when compared with fertile women.

Women with two gravidities form a remarkable group. We assume that their shorter lifespan is to be interpreted as a consequence of immunological reactions, for example, in terms of Rh incompatibility after the first pregnancy, which as a rule can be successfully carried to full term, but which leads to a higher risk of miscarriage with every additional gravidity. A high number of miscarriages probably subjects the organism to so much stress that this is reflected in a shortened lifespan. Interestingly enough, this same peculiarity is found in Bideau's (1986) data.

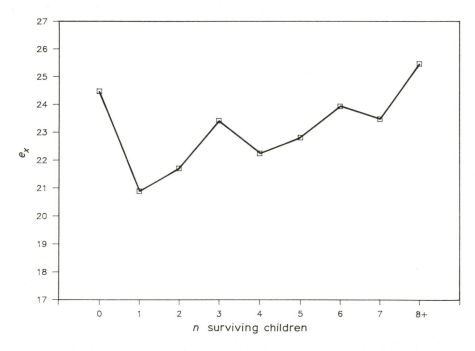

Figure 11.2 Future life expectancy of 47-year-old women (e_x) by number of surviving children (15 years and older).

The finding that is perhaps of most interest in terms of our topic is, however, the positive correlation between the number of surviving children and life expectancy (Figure 11.2). Three hypotheses come to mind when attempting to interpret this correlation.

The first one emphasizes the synchronous influence that the general living conditions exert on both the number of surviving children and longevity. Favourable living circumstances which enable one to raise many children to adulthood are simply less traumatic and therefore more likely to preserve life than those difficult living conditions, that do not permit many children to be raised to maturity.

A second psychological hypothesis sees longevity as the effect of mental and physical well-being (Lehr and Schmitz-Scherzer, 1976), which is more readily attained, the more children survive, because after having fulfilled the cultural norms and one's personal goals, the personal assessment of one's own life is likely to be positive (Rempel, 1985). In turn, this has the effect of preserving one's health and thus extending one's life (Reker and Wong, 1985).

A third hypothesis suggests that the number of surviving children and the lifespan of their mothers correlate with each other, because both variables are equally dependent on a third variable, namely on the age of the mother at the last birth of a child.

This hypothesis can definitely be substantiated. Fitness maximization requires not only that the number of children is adjusted to the prevailing socio-ecological conditions, but also the advantageous social placement of one's offspring by means of their socialization, marriage endowment with property, etc. In other words, fitness maximization also requires the allocation of the best living chances possible for the children because, on an average, optimal social placement translates – by whatever proximate mechanisms – into parental fitness gains. Parental patterns of behaviour, by which the fate of the children is influenced for the purpose of parental fitness maximization, can be designated as 'parental manipulation' (Alexander, 1974; Trivers, 1974) or as 'reproductive management' (Dickemann, 1984), irrespective of whether they are to the child's advantage or not. This means that the phase of postreproductive parenthood acquires a special importance in terms of life strategy, because now genetic reproduction is no longer in the foreground of parental efforts for fitness maximization but reproductive management is.

As death terminates the possibility of exerting a genetically selfish influence on the lives of children, the age of death should be a dependent function of the postreproductive lifespan. If the last birth is virtually considered to be the beginning of a 'countdown' of the decreasing possibility of manipulating the fate of one's children as they grow older, then – other things being equal – a positive correlation between the age at the last birth and the age at death is to be expected.

11.3.2 Longevity and reproductive management

The correlation between age at death (47 years and older) and the age at the last birth, or the last birth of a surviving child, respectively, are contained in Table 11.1. Figure 11.3 illustrates the increase of the further life expectancy of 47 year-old women with their age at the birth of the youngest surviving child.

These findings confirm our hypothesis of a positive correlation between longevity and age at last birth of a surviving child. Although the correlation is weak, it is statistically significant. This has nothing to do with any trivial correlation by virtue of there being common periods in life histories. We see the assumption that reproductive management plays a role as a variable of influence, as being confirmed by the finding in Table 11.1, namely that the correlation to (ii) is higher than to (i). Consequently, the existence of a living child postpones the time of death. The curve in Figure 11.3 also confirms a 'monotonous trend' ($r_s = 0.9429$, $p \leq 0.005$) in the increase in the age at death with the age at the last birth of a surviving child. This means, after all, an average difference of about four more years of life or not.

In a preindustrial demographic system with a comparatively high marital fertility, being older at the birth of the last surviving child frequently coincides with a higher number of children in general. Both variables are confounded. In our sample, the correlation between the age at the last birth and the number of surviving children amounts to 0.5363 $n = 751$; $p < 0.0001$). We performed a two-factor variance analysis in order to estimate the specific influence of both variables on the age at death. It appears that a main effect emanates from the age at the last birth of a surviving child (ANOVA, $F = 3.033$; $DF = 2$; $p = 0.049$), but there is no effect from the number of children born ($F = 1.222$; $DF = 3$; $p = 0.301$). When taking this result into account, the correlation documented in Figure 11.2 can be regarded as being a side-effect of another relationship that we presume is ultimate, namely the correl-

Table 11.1

The correlation between the age at death (47+) and the age at the last birth of a child (i), or the age at the last birth of a surviving child (ii)

Correlation	Number of of cases	Pearsons correlation coefficient	Statistical significance
(i) Age at death/age at the last birth of a child	752	0.06980	0.02787
(ii) Age at birth/age at the last birth of a surviving child	737	0.10597	0.00199

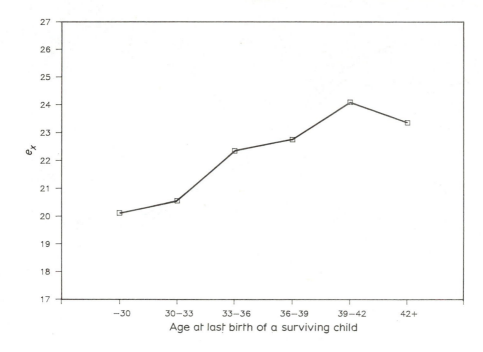

Figure 11.3 Future life expectancy of 47-year-old women (e_x) by their age at the last birth of a surviving child.

ation between the age at the last birth and the age at death. We therefore assume that the positive correlation between fertility and longevity that has been repeatedly reported in the literature, is, at least partially, an artefact, because the age at the last birth was not checked as a variable against longevity.

The possibility of parental manipulation exists to any noteworthy degree only as long as the social placement of the children has not been finished. In the premodern peasant society of Ostfriesland, the children's marriage probably marked a decisive shift in their living chances. The decisions on age at marriage (starting with own reproduction), selection of a mate, and the choice of an occupation, which are all relevant in terms of reproductive strategies, have been made by the time the children marry and settle down. Parental intervention in these essential aspects of life is no longer possible or necessary. And so to a certain extent, one ultimate reason for surviving during the phase of postreproductive parenthood is lost when the last child has married.

Figure 11.4 provides circumstantial evidence for the assumption that life

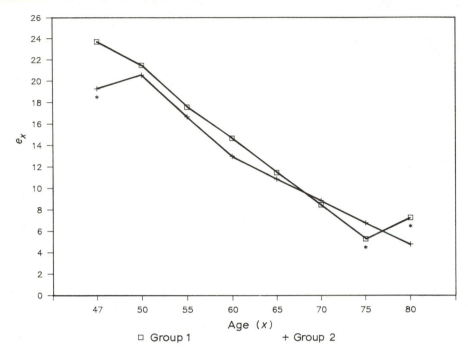

Figure 11.4 Future life expectancy (e_x) in two groups. Group 1: all x-year-old women, who have at last one unmarried child; group 2: all x-year-old women, whose children all are married or dead respectively.
*Values that are based on a value of n below 15 and are therefore rather uncertain.

expectancy really does depend on the possibility of exerting influence on a child's life, even if only to a slight degree. Depending on whether at least one child is still unmarried (group 1) or whether all of the children are either married or dead (group 2), the further life expectancy (e_x) of women aged 47 years and older shows a higher value for the women in group 1 (childless women are excluded).

There should not be any difference between both groups in the high age classes, because the unmarried children are now reaching an age which makes the taking up of reproductive activity improbable. Therefore, if the residual reproductive value of a child tends to drop to zero, supporting parental manipulation loses its fitness-maximizing significance. Accordingly, both curves should converge in the higher age classes. Unfortunately, the data with reference to the number of women aged 75 and older having unmarried children thins out, so that the convergence of both curves in these age classes can not be secured. We see our hypothesis as being supported up to the age class of the 65 year-olds, i.e. that the existence of at least one unmarried child has the effect of lengthening life.

11.4 CONCLUSION

The results allow the conclusion that reproductive activity can cause both phenotypic costs and phenotypic benefits for women.

The observation that genetic reproduction in general reduced the life expectancy somewhat in the sample under investigation, because married but childless women usually lived a little longer than mothers, belongs on the costs side of the balance. This applies to women in the postreproductive lifespans (47 years and older). In the premenopausal phase there were, of course, other costs that could not be taken into consideration in this study. These mainly include the life-threatening risks of giving birth (1.52% maternal mortality, Imhof, 1981) as survival costs, and the risks of birth traumata that lead to subsequent sterility, as fecundity costs.

Reproduction can, however, also have the effect of extending life; in other words, it is associated with a phenotypic benefit. If, the last delivery occurred during the last third of the fecund lifespan, the mothers lived about four years longer, on average, than if they had already previously completed their child-bearing. In our opinion, this is explained by the fact that mothers intervene in the lives of their children through behavioural patterns that can be called parental manipulation or reproductive management, and that they do this until the social placement of their children is completed. As this ability has the effect of maximizing inclusive fitness, it is strengthened by natural selection in a positive way and has become an essential part of life during the post-reproductive phase. Herein lies an approach to an evolutionary explanation of senescence (Hamilton, 1966). The earlier the last child was born, the earlier the mother's possibilities for genetically selfish interventions are exhausted and the weaker the positive selection pressure for the preservation of life in the postreproductive phase.

For humans, as for all organisms with reproductive management and parental manipulation the cost-benefit analysis of reproductive activity is not only a question of the physiological fecundity and survival costs; the life-lengthening effects resulting from reproductive management must also be taken into account when preparing the balance sheet.

11.5 SUMMARY

The variance in the age at death of postmenopausal women is explained, to a slight, but statistically significant degree, by the age of the women at the last birth of a surviving child. The later women complete their genetic reproduction, the longer they live. We regard the adaptive possibility of intervening in the fate of one's children during the postreproductive phase as the ultimate cause responsible for the correlation (i.e. parental manipulation, reproductive management). The earlier the children are placed socially, the earlier one chance of maximizing fitness, and thus, one ultimate reason for personal survival, has vanished.

REFERENCES

Alexander, R.D. (1974) The evolution of social behavior. *Ann. Rev. Ecol. System.*, **5**, 325–83.

Arvay, A and Takàcs, I. (1966) The effect of reproductive activity on biological ageing in the light of animal experiment results and demographical data. *Geront. Clin.*, **8**, 36–43.

Bell, G. (1984a) Measuring the cost of reproduction I. The correlation structure of the life table of a plankton Rotifer. *Evolution*, **38**, 300–13.

Bell, G. (1984b) Measuring the cost of reproduction II. The correlation structure of the life tables of five freshwater invertebrates. *Evolution*, **38**, 314–26.

Bideau, A. (1986) Fécondité et mortalité après 45 ans – L'apport des recherches en démographie historique. *Population*, **41**, 59–72.

Bryant, D.M. (1979) Reproductive costs in the house martin (*Delichon urbica*). *J. Anim. Ecol.*, **48**, 655–75.

Calow, P. (1979) The cost of reproduction – a physiological approach. *Biol. Rev.*, **54**, 23–40.

Clutton-Brock, T.H., Guinness, F.E. and Albon, S.D. (1983) The costs of reproduction to red deer hinds. *J. Anim. Ecol.*, **52**, 367–83.

Daly, M. and Wilson M. (1983) *Sex, Evolution, and Behavior*, 2 edn, Willard Grant, Boston.

Dickemann, M. (1984) Concepts and classification in the study of human infanticide: sectional introduction and some cautionary notes. In: *Infanticide – Comparative and Evolutionary Perspectives*, (eds G. Hausfater and S.B. Hrdy), Aldine, New York. pp. 427–37.

Fedigan, L.M., Fedigan, L., Gouzoules, S., Gouzoules, H. and Koyama, N. (1986) Lifetime reproductive success in female Japanese macaques. *Folia primatol.*, **47**, 143–57.

Gadgill, M. and Bossert, W.H. (1970) Life historical consequences of natural selection. *Am. Natur.*, **104**, 1–24.

Galef Jr., B.G. (1983) Costs and benefits of mammalian reproduction. In: *Symbiosis in Parent-Offspring Interactions*, (eds L.A. Rosenblum and H. Moltz), Plenum Press, New York and London, pp. 249–78.

Hamilton, W.D. (1966) The moulding of senescence by natural selection. *J. Theoret. Biol.*, **12**, 12–45.

Hirschfield, M.F. and Tinkle, D.W. (1975) Natural selection and the evolution of reproductive effort. *Proc. Natl. Acad. Sci. USA*, **72**, 2227–31.

Högstedt, G. (1981) Should there be a positive or negative correlation between survival of adults in a bird population and their clutch size? *Am. Natur.*, **118**, 568–71.

Imhof, A.E. (1981) Unterschiedliche Säuglingssterblichkeit in Deutschland, 18. bis 20. Jahrhundert – Warum? *Z. Bevölkerungswiss.*, **7**, 343–82.

Lawlor, L.R. (1976) Molting, growth and reproductive strategies in the terrestrial isopod, *Armadillidium vulgare. Ecology*, **57**, 1179–94.

References

Lawrence, P.S. (1941) The sex ratio, fertility, and ancestral longevity *Quart. Rev. Biol.*, **16**, 35–71.

Lehr, U. and Schmitz-Scherzer, R. (1976) Survivors and nonsurvivors – two fundamental patterns of aging. In: *Patterns of Aging – Findings from the Bonn Longitudinal Study of Aging* (ed. H. Thomae), Karger, Basel. pp. 137–46.

Ostfriesische Landschaft (ed.) (1966-1984), *Ostfrieslands Ortssippenbücher*, Bde. 4,6,8,9,14,15,16. Aurich, Ostfriesische Landschaft.

Powys, A.O. (1905) Data for the problem of evolution in Man. On fertility, duration of life and reproductive selection. *Biometrika*, **4**, 233–85.

Reker, G.T. and Wong, P.T.P. (1985) Personal optimism, physical and mental health – the triumph of successful aging. In: *Cognition, Stress, and Aging*, (eds. J.E. Birren and J. Livingston), Prentice-Hall, Englewood Cliffs, pp. 134–73.

Rempel, J. (1985) Childless elderly: what are they missing? *J. Marr. Fam.*, **47**, 343–8.

Ricklefs, R.E. (1977) On the evolution of reproductive strategies in birds: reproductive effort. *Am. Natur.*, **111**, 534–78.

Sachs, L. (1978) *Angewandte Statistik – Statistische Methoden und ihre Anwendungen.* Springer, Berlin, Heidelberg and New York.

Shine, R. (1980) 'Costs' of reproduction in reptiles. *Oecologia (Berl.)*, **46**, 92–100.

Tinkle, D.W. (1969) The concept of reproductive effort and its relation to the evolution of life histories of lizards. *Am. Natur.*, **103**, 501–16.

Trivers, R.L. (1974) Parent–offspring conflict. *Am. Zool.*, **14**, 249–64.

Voland, E. and Engel, C. (1986) Ist das postmenopausale Sterbealter Variable einer fitnessmaximierenden Reproduktionsstrategie? *Anthrop. Anz.*, **44**, 19–34.

Williams. G.C. (1966) Natural selection, the costs of reproduction, and a refinement of Lack's Principle. *Am. Natur.*, **100**, 687–90.

Wyshak, G. (1978) Fertility and longevity in twins, sibs, and parents of twins. *Soc. Biol.*, **25**, 315–30.

CHAPTER TWELVE

Household composition and female reproductive strategies in a Trinidadian village

Mark V. Flinn

'In effect a parent may dramatically increase the parental care available to its grandchildren by adding parents in the form of non-breeding offspring. A parallel circumstance may exist in jays and other group living birds with helpers at the nest . . . and in packs of canines dependent upon large game . . . in which frequently only one or two females breed while numerous individuals (older offspring? siblings?) share in parental duties.'

R.D. Alexander, 1974)

'Conflicting interests of co-wives in the children each bears . . . provide a major source of division . . . within households.'

(R.M. Keesing, 1976)

12.1 INTRODUCTION

Humans tend to reside with close kin (Murdock, 1967; Chagnon, 1979). Co-residence facilitates cooperation in a wide range of activities, such as defence, from predators and conspecifics, subsistence, and child care (Isaacs, 1978; Hames, 1979; Weisner, 1982; Chisholm, 1983; Turke, 1985). Human households, however, are not nepotistic nirvanas; competition may be equally pervasive (LeVine, 1962; Trivers, 1974; Wasser and Barash, 1983; Daly and Wilson, 1981, 1987).

In this chapter behavioural, demographic, and residential data from field-work in a Trinidadian village are analysed to test for associations among household composition, kin interactions, and the reproductive success of females. The objective is to examine the possible adaptive significance of cooperation and competition among co-resident female kin. The data indicate that: (1) only one female reproduces per household, suggesting that there is residential 'management' of reproduction, possibly including social suppression of the reproduction of other co-resident females, (2) non-reproducing females help their co-resident reproducing female kin by assisting with childcare and other household activities, and (3) the frequency of agonistic interactions between co-resident parents and daughters is highest when both mother and daughter are of reproductive age. These results suggest that evolutionary models of cooperative breeding systems (Brown, 1987; Emlen, 1984) may be helpful for understanding patterns of human reproduction (Lovejoy, 1981).

12.2 THEORY

Evolutionary theory provides useful models for analysing cooperative and competitive interactions among kin. Under appropriate circumstances, individuals may increase their genetic representation in future generations by helping relatives (Hamilton, 1964). Hence offspring may sometimes gain genetically by delaying or foregoing reproduction in order to help their parents reproduce, and vice versa. However, because individuals are genetically unique (with the exception of identical twins), conflicts of interest, and hence competition, are expected (Hamilton, 1964: Trivers, 1974: Alexander, 1974, 1979, 1987; West-Eberhard, 1983).

For example, honey bee workers co-operate (even to the point of suicidal stinging) to help their mother or sister reproduce. On the other hand, queens rigidly suppress reproductive attempts by workers, and newly emerged queens compete to the death with their sister-queens for possession of the hive and its consequent reproductive benefits (Wilson, 1971).

A number of social species have such 'cooperative breeding' systems: for example dwarf mongooses (Rood, 1978, 1986), African elephants (Dublin, 1983; Lee, 1987), naked mole rats (Jarvis, 1981), silver backed jackals (Moehlman 1983, 1986), termites (Brockmann, 1984), acorn woodpeckers (Stacey, 1979), Florida scrub jays (Woolfenden, 1975, 1981), and Galapagos mockingbirds (Curry, 1987). For reviews see: Alexander and Noonan, 1987; Andersson, 1984; Brockmann, 1984; Brown, 1983; Emlen, 1984; Reidman, 1982; and Skutch, 1987.

Co-operative breeding systems exhibit three general characteristics (Alexander and Noonan, 1989): (1) the lifetimes of the 'reproducer' and the 'helper' (usually parent and offspring) overlap; (2) 'helpers' are able to provide useful assistance to close genetic relatives; and (3) the alternatives to helping are unlikely to result in successful reproduction (e.g., because of limited resources such as nesting sites: Brown, 1987; Woolfenden, 1975; Emlen 1982a,b, 1984).

12.2.1 Reproductive co-operation in human families

Although human families are not eusocial in the honey bee sense (Wilson, 1975), the potential for reproductively important cooperative and competitive interactions clearly exists. The conditions necessary for co-operative breeding appear to be present in most human societies (Alexander, 1974; Quiatt, 1986; Strassmann 1986; Turke 1988).

First, the lifetimes of human parents and offspring overlap substantially. For example, during my fieldwork in Trinidad I resided in a household that included four generations (a 52 year-old woman, her 33 year-old daughter, a 19 year-old niece, four grandchildren aged 5–17 years old, and a newborn great-grandchild). Although extended families of this sort are not uncommon in human societies, overlap between adjacent generations (i.e., parent–offspring) is probably most significant.

Second, human juvenile and adolescent offspring on the one hand, and grandparents on the other, usually are capable of 'helping' with subsistence and childcare activities in most societies. Older children commonly 'babysit,' feed, and transport their younger siblings (Draper, 1977; Hames, 1988; Lamb and Sutton-Smith, 1982; Borgerhoff Mulder and Milton, 1985; LeVine, 1977; Munroe and Munroe, 1971; Super, 1981; Tronick *et al.*, 1987; Turke, 1988; Whiting and Whiting, 1975; Weisner and Gallimore, 1977; Wiesner, 1982). Similarly, grandparents frequently provide important assistance to their grand-offspring; indeed, menopause may have evolved because older women could increase inclusive fitness more by helping their offspring to reproduce than by continuing to attempt to reproduce themselves (Williams, 1966; Alexander, 1974; Mayer, 1981).

Finally, there are life-history periods during which individuals are capable of helping, but are incapable of reproducing on their own. Juvenile offspring and post-menopausal women are obvious examples. Adolescent and adult offspring also may have restricted reproductive opportunities because the social and economic resources necessary for successful reproduction, e.g. a farm, or bride-price, appear limited in most societies, and are frequently controlled (and inherited) by the family or some larger kin group (Boone, 1986; Flinn, 1986; Voland, 1984; Draper and Harpending, 1987; Lancaster and Lancaster, 1987). Even in those societies with an apparent abundance of physical resources, parental arrangement of mating relationships (marriages) may restrict offspring independence (Dickemann, 1981; Flinn and Low, 1986; Flinn, 1988a).

12.2.2 Reproductive competition in human families

About 50% of human pregnancies fail between conception and parturition (Shepard and Fantel, 1979). Wasser and Barash (1983) hypothesize that female–female competition in co-operative breeding systems may be an important cause of reproductive failure: '. . . where the conditions for survival of offspring are a function of the reproduction and support of other group members . . . some females may be able to improve current conditions for reproduction by suppressing the reproduction of others' (Wasser and Barash, 1983; see also Abbott *et al.*, 1986).

The conditions under which 'social suppression' of the reproduction of co-resident females could be adaptive are as follows: (1) resources within the household are limited, (2) the long-term reproductive costs to genetic relatives of suppression are not great, (3) suppressed females provide useful assistance, and (4) there is some mechanism (e.g. physiological, behavioural, and/or pheromonal) whereby suppression could occur.

In this chapter I consider three basic questions concerning reproductive co-operation and competition in a human population: (1) does household composition affect reproduction? (2) do co-resident non-reproductives 'help?' and (3) do females 'socially suppress' the reproduction of other co-resident females? Several types of information are useful for answering these questions. First, we need residence data. Second, we need reproductive histories. And third, we need information about the patterns of 'helping' behaviour and 'competitive' behaviour. The methods and field techniques utilized to gather these data are discussed below.

12.3 METHODS

During July and August, 1978, and October, 1979 through to April, 1980, I conducted field research in the village of Grande Anse and surrounding areas in rural North-central Trinidad (for description of the field site, see Flinn, 1988b). Some familiarity with Caribbean culture had previously been gained by part-time residence in the lesser Antilles during 1967-75. Three types of data were gathered that are useful for testing hypotheses about reproductive co-operation and competition: (1) residence, (2) genealogies, and (3) behavioural observations.

12.3.1 Residence and household composition

Information about where individuals resided in the village was obtained by interview and corroborated by incidental observation. Some individuals had multiple residences. For example, several children stayed with their grandparents or other relatives while their parent(s) were away working outside the village. And some mating relationships involved part-time cohabitation (and, hence, part-time parenting). Co-resident dyads are defined as those that sleep in the same house together more than 90% of the time and regularly eat meals in the same house. Households are defined as spatially separate residential houses (cf. Gonzalez, 1969).

At the time of the fieldwork, there were 112 households in the village. Fifty-two of these households included mother/offspring and/or grandparent/grandoffspring dyads. Thirty-one of these households included women that were pregnant or with an infant less than four years old. Twelve households included multiple generation maternal families, i.e. an adult female (ages ranged from 18 to 35) living with her parents and her offspring. Two households included multiple generation paternal families, i.e. an adult male living with his parents and his offspring. No multiple generation households had a co-resident step-parent or step-grandparent. Two households had grandparents/grandoffspring dyads but no parent/offspring dyads, and one household had grandmother/grandoffspring dyads but no parent/offspring dyads. There were no cases of co-residential polygamy.

There were 88 co-residential mother/offspring dyads, 17 part-time co-residential mother/offspring dyads, and 36 non-residential mother/offspring dyads in the village. There were no co-residential step-mother/step-offspring dyads.

Ten households had two or more co-resident women of reproductive age (17–40), involving a total of 18 dyads. All of these women were genetic relatives: thirteen mother/daughter dyads, three sister/sister dyads, one aunt/niece dyad, and one cousin/cousin dyad.

About 41% (26 out of 63) of the young adults (females aged 18–28 years old, males aged 20–30 years old) had both genetic-parents living in the village, 43% (27 of 63) had one genetic-parent living in the village, and 16% had no genetic-parents living in the village. These figures agree with other studies of rural Caribbean populations (Clarke, 1957; Otterbein, 1966; Smith, 1962). In a previous study, residence of parents in the village was found to be an important correlate of reproductive success (Flinn, 1986).

12.3.2 Genealogical information

Genealogical information is used to determine the age of individuals and the identities of relatives and non-relatives.

To collect genealogies I interviewed informants from each household, usually adult females, obtaining the names, genealogical relationships, ages, and current residences of all the relatives (blood and affinal) that they could remember (Chagnon, 1974). These interviews were well received. Upon returning from the interview, I assigned unique I.D. numbers to each individual collected in the genealogies and put all of the above information on index cards for each individual. Discrepancies and questionable paternity assignments were checked by additional interviews. I found it useful to seek redundant information from several informants to allow for cross-checking

(Chagnon, 1974). The genealogical information was analysed by computer (in the US) for cross reference with the behavioural and residence data.

Because age is an important determinant of reproduction, an age-specific fertility curve was estimated from all births since 1950 to women (of known ages) residing in the village. The age-specific fertility rate was estimated by dividing the number of births by women of a given age by the total number of women in that age group (e.g. 19 of 93 women gave birth between the ages of 24 and 25, giving a rate of 0.204 births per woman year for that age).

Females between the ages of 17–40 (inclusive) are referred to as being of *reproductive age*. Females that were pregnant or were mothers of infants less than four years of age are referred to as *reproductive females*. Females of reproductive age that were not pregnant and were not mothers of infants less than four years of age are referred to as *non-reproductive females*.

12.3.3 Day-to-day behaviour

I collected data detailing the day-to-day behaviour of the villagers with a 'behavioral observation route instantaneous scan sample' procedure (Munroe and Munroe, 1971; Munroe *et al.*, 1983; Altmann, 1974; Johnson, 1975; Denham, 1978; Hames, 1979; Betzig and Turke, 1986). Scan data are useful for the study of social relationships because they can provide objective, quantitative measures of behaviour that would be difficult or unreliable to obtain by questionnaire techniques (Draper, 1977; Johnson, 1978; Blurton Jones, 1972; Gross, 1984; Betzig and Turke, 1986).

The behaviour scan procedure was as follows: I travelled a set 4.7 km route through the study site once or twice daily, starting at a randomly determined time and place on the route. The route went through the entire village, passing within 20 m of each inhabited house and each community structure (e.g. church, cricket field, water outlets). Because village houses are quite open, and because the route passed close by each house, observability was excellent. Each time an individual was observed, I recorded (with a notebook and/or tape recorder) the time, location, individual, and behaviour. This information was coded within 48 h onto computer format sheets. For each 'observation', the date, time, one of 1375 location codes, one of 480 individual identification numbers, and one or more of 475 behaviour codes were numerically recorded for computer analysis. For example, on March 7, 1980 at 6.25 am I observed Lucille W. in her household's kitchen making 'tea' (hot milk) with Charmaine W. (her 5 year-old daughter) at her side. Lucille was telling Charmaine how to mix up the powdered milk properly. Charmaine was sipping at a cup of the milk. This observation was coded:

date	time	location	individual	behaviour
070380	0625	0016	0013	304342
070380	0625	0016	0019	354

I recorded about 33 000 observations in the above fashion over a period of six months (173 scan routes on 152 days). Of these observations, 24 577 form the data base used in this chapter. I have excluded observations recorded during the first two weeks of the procedure, observations recorded during scan routes in which fewer than 50% of the villagers were observed, observations of visitors to the village and observations of unidentified individuals. For further description of the scan technique, see Flinn (1983, 1988a). Three measures of behaviour are computed from the scan data:

Frequency of behavioural interaction between ego and another individual was determined by dividing the number of times they were observed interacting with each other by the total number of observations of ego. This provides an estimate of the proportion of time that ego interacts with that particular individual.

Frequency of childcare was determined by dividing the number of observed 'childcare' interactions by the total number of observations of the individual. Child care interactions are defined as those behaviours that involved close physical proximity between the child (0–4 years old) and an older individual, and had some apparent benefit to the child. Thirteen of the 475 behaviour codes were included in this category: babysitting, washing child, changing diapers/clothes, miscellaneous grooming (e.g. brushing hair), feeding child, nursing child, holding child, teaching child, instructing child to do a task, singing to child, carrying child, playing with child, physical affection – tickling, rubbing, kissing child. Of the total of 24 577 observations of all villagers, 1917 (7.8%) involved child care interactions. As a frame of reference, I subjectively consider the frequency of childcare in Grande Anse families to be about equal to or slightly above that of American middle-class families. The behavioural data (section 12.4) show frequencies similar to those obtained by Weisner (1987) in a study of childcare among the Abaluyia of western Kenya. Because the age of a child may be an important determinant of child/caretaker interactions (Munroe and Munroe, 1971; Flinn, 1983), analyses using frequency of childcare were adjusted for age effects (Flinn, 1988c).

Frequency of agonistic behavioural interactions was determined by dividing the number of observed 'agonistic' interactions between ego and another individual by the total number of interactions between ego and that individual. Agonistic interactions are defined as those behaviours that involved physical or verbal combat (e.g. 'spanking' or 'arguing') or expressions of injury inflicted by another individual (e.g. 'screaming in pain or anguish' or 'crying'). Twenty-six of the 475 behaviour codes were included in this category (Flinn, 1988b). Of the total of 24 577 observations of all villagers, 1218 (4.8%) involved agonistic interactions. Most of these were verbal behaviours (92.5%). Of the total of 5343 observations of parent/offspring interactions, 318 (6%) were agonistic. Most of these were verbal behaviours (289, or 91%). As a frame of reference, I subjectively consider the level of agonism in Grande Anse families to be about equal to that of American middle-class families. Although agonistic interactions probably reflect a conflict of interests, some agonistic interactions might be nepotistic, or at least mutually beneficial in the long term. For example a mother might reprimand her daughter for a behaviour that could be dangerous, such as walking into the street, or pulling a pot from on top of the stove. From the scan data, it is difficult to distinguish nepotistic agonistic interactions from non-nepotistic agonistic interactions. It is assumed that the rate of agonistic interactions is a good indicator of the level (degree) of conflict between individuals (Flinn, 1988b). I was unaware of any instances of what would be called 'physical child abuse' i.e. serious injury inflicted on a child by a parent (Daly and Wilson, 1985) during the period of the field-work.

The above three measures provide objective estimates of behaviour. In section 12.4 these measures are used to test hypotheses concerning cooperative and competitive relationships among co-resident females. Unfortunately these measures are rather crude compared to the those used by developmental psychologists studying modern North American or European populations (Super, 1981), and less detailed than those used by anthropologists focussing explicitly on mother–infant interactions (Borgerhoff Mulder and Milton, 1985).

Note: The data analysed in this chapter are complete for each villager. Residence, genealogies of at least two generations in depth, and behaviour (scan sample) are known for each of the 342 individuals in the village population.

12.4 RESULTS AND DISCUSSION

The data are presented in three sections. First, household composition data and female reproductive success data are analysed to test whether co-residence affects reproduction. Second, behavioural data are analysed to test whether co-resident non-reproductive females function as 'helpers at the nest'. And third, behavioural data are analysed to test if reproductive females attempt to socially suppress reproduction by co-resident non-reproductive females.

12.4.1 Household composition and female reproductive success

During the period of the fieldwork (August 1978 to May 1980), 31 out of 106 households in the village had 'reproductive' females (pregnant or with an infant aged 0–3). Of these 31 households, ten households had more than one female of reproductive age (17–40). In six of these ten households mothers were reproductive, and daughters non-reproductive. Three of the households had a reproductive daughter, a non-reproductive mother, and one or more non-reproductive sisters. And one household had a reproductive daughter, a non-reproductive mother, a non-reproductive grandmother, and a non-reproductive aunt.

No female became pregnant while co-residing in a household with a 'reproductive' female, even though most of these 'non-reproductive' females ($n = 14$ of 18) were reported to be sexually active (12 of these 14, however, were not co-residing with their mates). The expected number of births by these women (computed from their age-specific fertility rates) in this 21 month time period was 2.93, a small enough number that the absence of births in this group could be due to chance. However, examination of residence histories and genealogies over the ten year period 1970–1980 (pre-1970 data were incomplete) also indicates that only one female per household reproduced (within a five year period). Here the expected number of births was 16.4 (vs 0.0 actual), a substantial difference. In brief, there was only one reproducing female per household.

Cooperative breeding theory predicts that females with 'helpers' should outreproduce females without 'helpers.' In an ingenious test of this hypothesis, Turke (1988) examined demographic data from an Ifaluk atoll (Yap state, Micronesia). Noting that females are more likely to be useful as helpers (because of matrilocal residence and high female contribution to subsistence and childcare in this population), Turke predicted that females with first-born daughters (helpers) should outreproduce females with first-born sons. The data supported this prediction (Turke, 1988). Grande Anse is similar to Ifaluk in that residence tends to be matrilocal, and females are the primary child caretakers. In Grande Anse, as in Ifaluk, daughters are more likely to be useful 'helpers' than are sons. However, the data from Grande Anse do not support the prediction of higher reproductive success for women with first-born daughters than for women with first-born sons (Figure 12.1).

The failure to replicate Turke's (1988) results does not necessarily falsify the general hypothesis that women with helpers outreproduce women without helpers. There may be differences in the importance of parity between the two populations, perhaps because of closer birth spacing in Grande Anse, and/or differences in inheritance patterns. A more appropriate prediction for the Grande Anse population is that reproductive females with co-resident non-reproductive females (potential helpers) have higher reproductive success

214

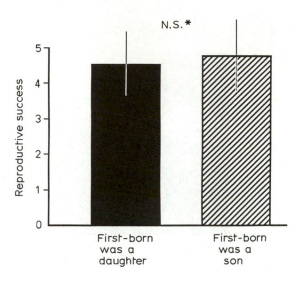

Figure 12.1 The reproductive success (number of offspring surviving to one year of age) of women with first-born daughters (black bar), vs. first-born sons (striped bar). The data indicate that women with first-born daughters ($n = 17$) did not have higher reproductive success than women with first-born sons ($n = 21$). Sample includes all women resident in Grande Anse in 1980 with two or more offspring. 95% confidence intervals (1.96 × Standard Error) for the means are indicated by the vertical lines. *N.S. = not significant.

than reproductive females without co-resident non-reproductive females. The data in Figure 12.2 support this prediction; however, this result could be due to larger families being more likely to have a daughter staying at home. It is not possible to determine the direction of causality from this result. Note that only females over forty years of age with two or more offspring are included in the sample to control for the fact that females with 'helpers' already have two offspring.

If daughters are postponing reproduction in order to help their mothers reproduce, then the presence or absence of younger siblings should affect the age at which daughters begin to reproduce. During the period 1970–1980, no female ($n = 14$) of reproductive age that lived with her mother became pregnant until her mother's last-born child was at least four years old. Females with co-resident siblings that were more than ten years younger had their first baby at a later age than other females (Figure 12.3).

Similarly, if mothers are ceasing reproduction early in order to help their daughters reproduce, then mothers with co-resident reproductive-aged daughters should cease reproduction sooner than mother's without co-resident reproductive-aged daughters. The data in Figure 12.4 support this prediction:

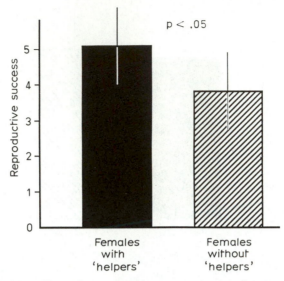

Figure 12.2 The effect of co-resident non-reproductive females on reproductive success (number of offspring surviving to one year of age). Reproductive females ($n = 9$) with co-resident non-reproductive females averaged significantly higher reproductive success than did reproductive females ($n = 29$) without co-resident non-reproductive females ($\bar{x} = 5.11$ vs. $\bar{x} = 3.79$, $t = 2.92$ $p < 0.05$). Sample includes all females resident in Grande Anse in 1980 with two or more offspring. 95% confidence intervals for the means are indicated by the vertical lines.

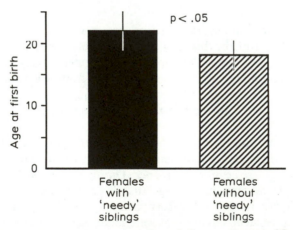

Figure 12.3 The effect of younger co-resident siblings on the age of first reproduction for females. Females ($n = 11$) that at age 18 were co-residing with younger (> 10 years age difference, i.e. 'needy') siblings averaged a significantly later age at first birth than did females ($n = 37$) that were not ($\bar{x} = 21.4$ vs. $\bar{x} = 18.7$, $t = 3.4$, $p < 0.05$). Sample includes all females resident in Grande Anse in 1980 with one or more offspring. 95% confidence intervals for the means are indicated by the vertical lines.

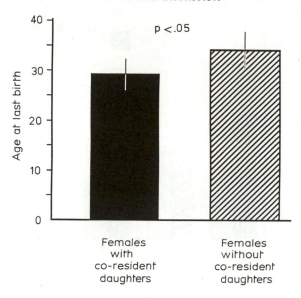

Figure 12.4 The effect of older co-resident daughters on mother's age of last birth. Mothers ($n = 12$) (black bars) at age 40 with co-resident daughters aged 18 or older ceased reproducing at a significantly earlier age than did other mothers ($n = 42$) (striped bars) ($\bar{x} = 29.6$ vs. $\bar{x} = 33.7$, $t = 3.22$, $p < 0.05$). Sample includes all mothers resident in Grande Anse that turned 40 during the years 1970-80 that had more than one offspring. 95% confidence intervals for the means are indicated by the vertical lines.

mothers that had, at age 40, reproductive-aged (> 17 years old; i.e. mother's age minus daughter's age is < 23) co-resident daughters, had ceased reproducing at earlier ages than did other mothers, but mothers with reproductive-aged *non*-resident daughters had not.

The data presented in Figures 12.1–12.4 suggest that there is suppression of reproductive overlap between mothers and daughters. Co-operative breeding theory predicts that reproductive females should be achieving higher reproductive success as a consequence (Brown and Brown 1972; Woolfenden, 1975) However, because a female may be both a non-reproductive helper at one period in her life history, and a reproductive female at another period, it is difficult to use lifetime reproductive success as a measure of the effects of co-residential cooperative breeding. Age-specific fertility is a more useful measure. Figure 12.5 illustrates the effect of co-residence on age-specific fertility of mothers.

The data presented in Figure 12.5 indicate a balance between the reproductive benefits and costs of co-residence to mother and daughter. Mothers with older co-resident daughters show an increased reproductive rate during the ages 27–30, but a decreased reproductive rate during the ages 35–42.

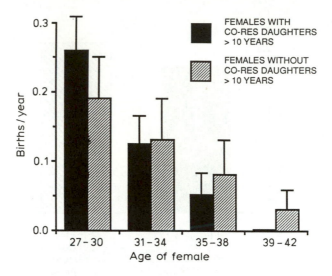

Figure 12.5 The effect of co-resident daughters aged 10 years and older on the age-specific fertility of their mothers. Mothers ($n = 73$ female/years) (black bars) with co-resident daughters aged 10 years and older have higher reproductive rates during ages 27–30, but lower reproductive rates during ages 35–42, than do other mothers ($n = 487$ woman/years) (striped bars). Sample includes all females resident in Grande Anse in 1980 with two or more offspring. 95% confidence intervals for the means are indicated by the vertical lines.

These results are consistent with the hypothesis that (a) help from non-reproductive co-resident daughters increases reproduction, but (b) helping reproductive co-resident daughters decreases reproduction.

Figure 12.6 illustrates the affect of co-residence on age-specific fertility of daughters. These data also indicate a balance between the reproductive benefits and costs of co-residence to mother and daughter. Daughters with co-resident mothers show a decreased reproductive rate during the ages 18–21, but a slight increase in reproductive rates during the ages 22–29 (note: the reproductive rates of co-resident daughters are somewhat reduced by the inclusion of three females aged 22, 25, and 27 that apparently were non-reproductive because of physical or mental handicaps). These results are consistent with the hypothesis that (a) helping reproductive co-resident mothers decreases reproduction, but (b) help from non-reproductive co-resident mothers increases reproduction.

The data presented in Figures 12.5 and 12.6 partially indicate the reproductive benefits and costs of co-residence for mothers and daughters. For example, mothers with older co-resident daughters show a maximum average increase of 0.06 offspring/year at age 29, and a maximum decrease of 0.064 offspring year at age 35. Daughters with co-resident mothers show a maxi-

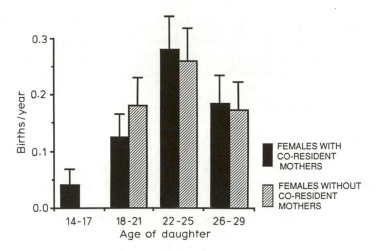

Figure 12.6 The effect of co-resident mothers on the age-specific fertility of their daughters. Females ($n = 56$ female/years) (black bars) with co-resident mothers have lower average reproductive rates during the ages 18–21 ($t = 4.64$, $p < 0.05$), but slightly increased (N.S.) reproductive rates during 22–29, than do other females ($n = 409$ female/years) (striped bars). 95% confidence intervals for the means are indicated by the vertical lines.

mum average increase of 0.036 offspring at age 26, and a maximum decrease of 0.12 offspring/year at age 20. Neither daughter nor mother would appear to increase inclusive fitness by delaying daughter's reproduction past the age of 18 (unless the daughter is gaining some benefit that will result in increased reproduction later on). Daughters that stay at home past the age of 18 or 19 to be non-reproductive 'helpers' are probably making the best of a bad situation. Helping mother may be the best option for daughters that have not yet established an acceptable mating relationship. Helping daughters to reproduce, however, may increase a mother's inclusive fitness once she has passed a certain age.

Several households included co-resident sisters of reproductive age. No female reproduced that was co-resident with a reproductive sister ($n = 6$ pairs and one triad). Several females reproduced (5 of 9) that had co-resident brothers (aged 20–30) without resident infants, but no females (aged 17–40) reproduced (0 of 5) that had co-resident brother's infants ($p = 0.086$, Fisher's exact test), suggesting that reproduction is inhibited by the co-residence of infants rather than the co-residence of a reproductive female *per se*.

The data presented in this section indicate that female reproduction is affected by household composition. Female life history strategies are evidently influenced by the relative ages of mother and daughter, and the age at which daughters move away from home. The data suggest that helping non-descendant kin (siblings, nephews and nieces, grandchildren) could be the reason for reproductive suppression. However, behavioural data are required to demonstrate that co-resident kin provide assistance. We also have not falsified the hypothesis that reproductive suppression results from coercion by the reproductive female. In the following sections I will examine, first the evidence for helping, and second, the evidence for coercion.

12.4.2 Behavioural evidence of 'helping'

In this section behavioural scan data are analysed to determine whether or not co-resident non-reproductive females are providing useful assistance (i.e. childcare) for the reproductive female.

Figure 12.7 compares the frequency of sibling childcare by females and males according to age. The data indicate that females provide substantially more sibling childcare than do males, particularly in the 11–25 age categories. This result is consistent with other gender studies of childcare (Barry and Paxton, 1971; Turke, 1985; Weisner, 1987) and supports the assumption that daughters are more important 'helpers at the nest' than are sons.

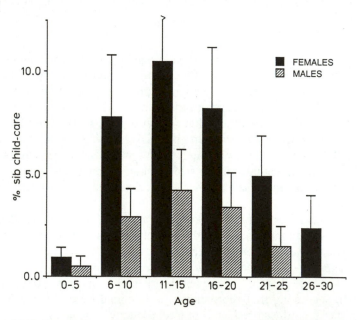

Figure 12.7 Comparison of the observed frequencies of sibling child-care by females and males. Females ($n = 43$) (black bars) are observed to provide more child-care to their siblings than males ($n = 51$) (striped bars) in all age categories. Sample includes all residents of Grande Anse with younger siblings under the age of ten. 95% confidence intervals for the means are indicated by the vertical lines.

Results and discussion

Childcare, however, is not the only contribution that offspring can make to the household. In many societies older offspring make important economic and/or defence contributions to the household (Irons, 1979; Chagnon, 1979, 1982; Strassmann, 1986) that could affect reproduction. In Grande Anse the difference between the material-monetary economic contributions of sons and daughters is negligible; both provide small amounts. Moreover, there is no correlation between household land ownership or income and female reproductive success (Flinn, 1983, 1986), so additional economic contributions by offspring apparently would not affect their mother's reproduction. Although it would seem that alleviating the mother from economic tasks would allow her to be more reproductive, alleviation from childcare evidently has a greater effect on reproduction (cf. Chisholm, 1981).

Figure 12.8 illustrates the overall frequencies of childcare by female siblings, mothers, and grandmothers. Although mothers have the highest frequencies of childcare, substantial amounts are provided by siblings and grandmothers. Non-reproductive female relatives appear to be an important resource for dependent children and their mothers.

To test the hypothesis that co-resident non-reproductive females alleviate some of the mother's childcare duties, Figure 12.9 compares the frequencies

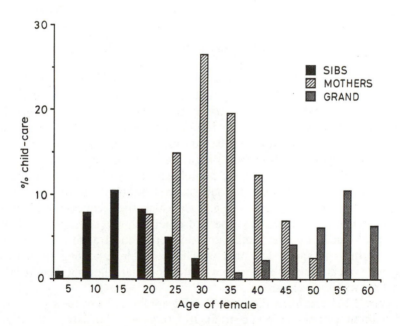

Figure 12.8 The observed frequencies of child-care by sisters (black bars), mothers (striped bars), and grandmothers (stippled bars) by five-year age categories.

221

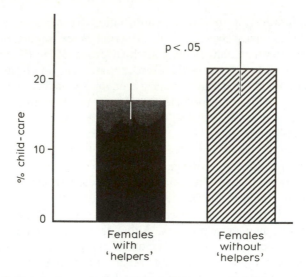

Figure 12.9 Comparison of the observed frequency of child-care by reproductive mothers with and without 'helpers' (co-resident non-reproductive females 10 years of age and older). Mothers ($n = 9$) with 'helpers' had a lower proportion of child-care observations than did mothers ($n = 19$) without 'helpers' ($\bar{x} = 17.8\%$ vs. $\bar{x} = 22.0\%$, Chi-square $= 4.1$, $p < 0.05$). Sample includes all mothers with co-resident offspring aged 0–4. 95% confidence intervals for the means are indicated by the vertical lines.

of childcare by mothers with and without co-resident non-reproductive females.

The data indicate that mothers with co-resident non-reproductive females have lower frequencies of childcare than do mothers without co-resident non-reproductive females, suggesting that mothers with 'helpers' have more free time to pursue other activities. This result is similar to Borgerhoff Mulder and Milton's (1985) finding in their study of infant care among the Kipsigi that mothers in polygynous households spend less time caring for their infants than do mothers in monogamous households. Both Grande Anse and the Kipsigi provide an interesting contrast to Chisholm's study of Navaho mother–infant relations (1981, 1983), in which the frequency of mother–infant interactions is higher in extended family camps (containing potential 'helpers') than in nuclear family units. Whereas in Grande Anse 'helpers' allow mothers to spend *less* time with their children, among the Navaho 'helpers' evidently free mothers from economic duties, allowing them to spend *more* time with their children.

Figure 12.10 examines the rates of childcare for children (0–4 years) with and without co-resident non-reproductive females. The data indicate that children with co-resident non-reproductive females receive slightly more care than do children without co-resident non-reproductive females. This result is

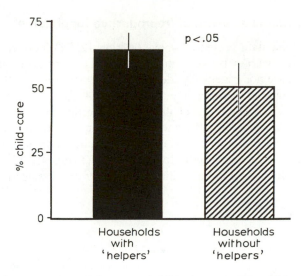

Figure 12.10 Comparison of the rates of childcare received by children 0–4 years of age in households with and without 'helpers' (co-resident non-reproductive females 10 years of age and older). Children in households with 'helpers' were observed in childcare interactions a higher proportion of the time than were children in households without 'helpers' ($\bar{x} = 64.2\%$ vs. $\bar{x} = 50.7\%$, Chi-square $= 5.7$, $p < 0.05$). Sample includes all children aged 0–4 resident in Grande Anse. 95% confidence intervals for the means are indicated by the vertical lines.

consistent with several previous studies of 'household density' (number of caretakers) and infant care (Chisholm, 1983; Munroe and Munroe, 1971; Weisner and Gallimore, 1977; Munroe and Munroe, 1984; Borgerhoff Mulder and Milton, 1985). The 'quality' of care received by Grand Anse children from mothers versus older sisters is uncertain; Borgerhoff Mulder and Milton's (1985) study suggests that the quality of care given by mothers and alternative caretakers is similar (cf. Konner, 1975; Leiderman and Leiderman, 1977), but that with increasing numbers of caretakers the quality of care may go down. The benefits and costs of alloparenting for the infant/child are unknown.

In summary, the behavioural data indicate that co-resident non-reproductive females do provide useful assistance (i.e. childcare) for the reproductive female, consistent with the hypothesis they are functioning as 'helpers at the nest'. It is not known whether 'helpers' derive some benefit from childcare, such as useful practice in preparation for their own future offspring (Hrdy, 1976, 1981; McKenna, 1981, 1987). Women suggested to me that child care was sometimes a chore, and that sibling caretakers expected some return benefit from the mother.

12.4.3 Behavioural evidence of 'reproductive suppression'

The demographic data presented above (section 12.4.1) indicate suppression of reproductive overlap between mothers and daughters. Because mothers and daughters are not genetically identical, there may arise conflicts of interest over the timing of reproductive (Trivers, 1974; Alexander, 1974; Parker and McNair, 1979). Figure 12.11 illustrates the estimated period of parent–offspring conflict concerning the age at which daughter begins reproduction, and the age at which mother ceases reproduction, based on a smoothed curve (least squares) of the age-specific fertility rates (cf. Figures 12.5 and 12.6). For example, the age specific fertility for 18 year-old females is 0.091, and

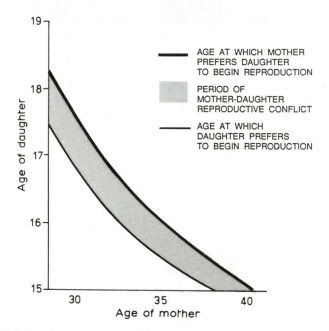

Figure 12.11 Estimation of period of co-residential mother–daughter reproductive conflict in monogynous households from age-specific fertility rates. Mothers share 50% of their genes identical by descent with their offspring, but 25% on average with their grand-offspring. Hence mothers are predicted to prefer to remain reproductive (keeping their co-resident daughters non-reproductive) up until the daughter's potential reproductive rate (estimated from age-specific fertility smoothed curve) is twice that of the mother (thick line). Daughters, on the other hand, share 50% of their genes with their own offspring and on average 50% (full sib) or 25% (half-sib) with their mother's offspring. Hence daughters will prefer to begin reproducing (causing cessation of mother's reproduction) while their potential reproductive rate is slightly less than that of their mother (depending on the frequency of half-sibs; figure is based on 25% half-sib rate, or 43.75% average relatedness with mother's offspring) (thin line). A period of conflict (stippled area) exists between the two curves.

that for 36 year-old females is 0.060. Because mothers are related by 50% to their own offspring, but by 25% on average to their daughter's offspring, we might expect that a 36 year-old mother with an 18 year-old daughter would prefer to remain reproductive ($0.5 \times 0.060 > 0.25 \times 0.091$); the daughter, on the other hand, would prefer to begin reproduction ($0.5 \times 0.091 > 0.5 \times 0.060$). The degree of conflict will be increased by multiple paternity (see caption, Figure 12.11).

If reproductive suppression occurs due to coercion by the reproductive female, then we expect agonistic interactions to be most frequent during the period in which there is a conflict of interests (Wasser and Barash, 1983). Figure 12.11 suggests that mother–daughter reproductive conflict is most intense between daughters >17 years-old and mothers <38 years-old. Figure 12.12 compares the frequencies of agonistic interactions between parents and daughters according to the age of the daughter (four-year age categories) and whether the co-resident mother is reproductive (father–daughter interactions are included because the interests of fathers and mothers are identical in this regard; the patterns of mother–daughter and father–daughter interactions are similar).

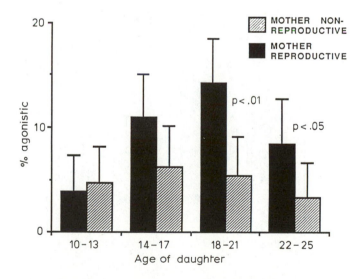

Figure 12.12 Comparison of the proportions of agonistic interactions between parents and daughters in households with reproductive mothers (black bars) and households with non-reproductive mothers (striped bars). Daughters aged 18–25 have higher average proportions of agonistic interactions with their parents if their mother is reproductive than if their mother is non-reproductive. Sample includes all households with co-resident parent(s) and daughters aged 10–25. 95% confidence intervals for the means are indicated by the vertical lines.

The data indicate significantly higher frequencies of agonistic interactions between parents and 18–21 year-old daughters when mothers are reproductive, consistent with the mother–daughter reproductive conflict hypothesis.

Part of this increased agonism with reproductive-aged daughters could be due to efforts by the parents to control the mating activities of the daughter (Flinn, 1988a). The 'daughter guarding' hypothesis predicts a high frequency of agonistic interactions with reproductive-aged daughters regardless of mother's age. The mother–daughter reproductive conflict hypothesis predicts an increased rate of agonistic interactions with younger, reproductive-aged mothers. Figure 12.13 compares the frequencies of agonistic interaction between parents and daughters according to the age of the mother (four-year age categories) and whether the co-resident daughter (of reproductive age) is reproductive.

The data indicate significantly higher frequencies of agonistic interactions between parents and daughters when mothers are < 38 and co-resident reproductive-aged daughters are non-reproductive, consistent with the

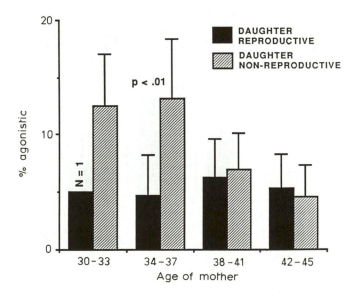

Figure 12.13 Comparison of the proportions of agonistic interactions between parents and daughters in households with reproductive daughters (black bars) and households with non-reproductive daughters (striped bars). Mothers aged 30–37 have higher average proportions of agonistic interactions with their daughters if the daughter is non-reproductive than if the daughter is reproductive. Sample includes all households with co-resident daughters aged 10–25 and mothers aged 30–45. 95% confidence intervals for the means are indicated by the vertical lines.

mother–daughter reproductive conflict hypothesis. In general, the highest frequency of agonistic interactions between parents and daughters occurs in households with reproductive mothers and reproductive-aged daughters.

In summary, the behavioural data are consistent with the hypothesis that co-resident reproductive-aged females are reproductive competitors. Determining whether or not reproductive suppression occurs as a consequence of agonistic interactions (e.g. via stress-induced fertility reduction), however, requires additional research. The degree of conflict among unrelated co-wives in polygynous societies is expected to be much higher (Daly and Wilson, 1983; Le Vine, 1962).

12.5 SUMMARY AND CONCLUDING REMARKS

The data support three general conclusions: (1) there is reproductive management among co-resident females such that only one female reproduces per household, (2) non-reproductive females provide substantial amounts of child-care, thereby reducing the burden of reproductive females, and (3) there are high levels of agonistic interactions between co-resident females that are potentially reproductive. These findings are consistent with the hypothesis that household composition has important effects on female reproductive strategies in Grande Anse. Several important questions, however, remain unanswered.

First, it is difficult to determine the extent to which the apparent household monogyny results from an association between residence choice and reproductive planning (conscious or unconscious). Daughters that are likely to begin reproducing in the near future (i.e. who have established mating relationships) are probably more likely to move away from home than are daughters that are not, especially if their mothers are still reproductive. Clearly humans are not obligately monogynous; co-resident co-wives (often sisters) reproduce concurrently in many human societies (although several studies indicate reduced fertility among women in polygynous relationships – see Irons, 1983). The observed household monogyny in Grande Anse is not the result of reproductive suppression of the sort occurring among eusocial insects. Rather, it appears to be a flexible system that provides nepotistic opportunities for females that are non-reproductive for other reasons (age, lack of suitable mating relationship). Caring for the dependent children of a close female relative evidently provides a sufficient inclusive fitness maximizing option for non-reproductive females that their interest in direct reproduction is diminished (hence the later age at first birth for females with dependent sibs – Figure 12.3). Reproductive females (and their mates), however, may exploit the nepotistic interests of co-resident non-reproductive females, sometimes generating conflicts of interest (Figure 12.12).

Second, the mechanisms underlying the hypothesized female reproductive strategies remain obscure. We do not know how females communicate (and hence potentially manipulate) their reproductive and nepotistic intentions. We do not know what cues females use to make residence choice decisions. We do not know what behavioural or physiological mechanisms might be responsible for reproductive suppression (perhaps there are connections with related phenomena such as concealment of ovulation or menstrual synchrony – Alexander and Noonan, 1979; McClintock, 1971). We do not know whether males influence female strategies. Finally, we do not know whether the Grande Anse environment differs from the environment of human evolutionary adaptedness in ways that render current behaviour non-adaptive. Additional research is required to address these issues.

ACKNOWLEDGEMENTS

I would like to thank Richard Alexander, Lars Rodseth, Beverly Strassmann, and Paul Turke for discussion and unpublished manuscripts.

REFERENCES

Abbott, D.H., Keverne, E.B., Moore, G.F. and Yodyingyuad, U. (1986) Social suppression of reproduction in subordinate talapoin monkeys, *Miopithecus talapoin.* In: *Primate Ontogeny, Cognition, and Social Behavior* (eds J.G. Else and P.C. Lee), Cambridge University Press, Cambridge, pp. 329–42.

Alexander, R.D. (1974) The evolution of social behavior. *A. Rev. Ecol. Systematics,* **5**: 325–83.

Alexander, R.D. (1979) *Darwinism and Human Affairs.* University of Washington Press, Seattle.

Alexander, R.D. (1987) *The Biology of Moral Systems.* Aldine Press, Hawthorne, NY.

Alexander, R.D. and Noonan, K. (1979) Concealment of ovulation, parental care, and human social evolution In: *Evolutionary Biology and Human Social Behavior* (eds N. Chagnon, and W. Irons), Duxbury Press, North Scituate, Mass., pp. 436–53.

Alexander R.D. and Noonan, K. (1987) The evolution of eusociality. In: *The Natural History and Social Behavior of Naked Mole Rats* (eds P.W. Sherman, J. Jarvis and R.D. Alexander), (in press).

Altmann, J. (1974) Observational study of behavior: sampling methods. *Behaviour,* **49**, 227–66.

Andersson, M. (1984) The evolution of eusociality. *A. Rev. Ecol. Systematics,* **15**, 165–189.

Axelrod, R. and Hamilton, W.D. (1981) The evolution of cooperation. *Science,* **211**, 1390–6.

Barry, H. III and Paxton, L.M. (1971) Infancy and early childhood: cross cultural codes. *Ethnology,* **10**, 466–508.

Betzig, L. and Turke, P. (1986) Parental investment by sex on Ifaluk. *Ethcology and Sociobiology,* **7**, 29–37.

References

Blake, J. (1961) *Family Structure in Jamaica.* Free Press, New York.

Blurton Jones, N. (1972) (ed.) *Ethological Studies of Child Behavior.* Cambridge University Press, Cambridge.

Boone, J. (1986) Parental investment and elite family structure in preindustrial states: a case study of late medieval–early modern Portugese genealogies. *Am. Anthrop.,* **88**, 859–78.

Borgerhoff Mulder, M.B. and Milton, M. (1985) Factors affecting infant care in the Kipsigis, *J. Anthrop. Res.,* **41** (3) 231–62.

Brockmann, H.J. (1984) The evolution of social behaviors in insects. In: *Behavioral Ecology,* 2nd edn, (eds J.R. Krebs and N.B. Davies), Sinauer Associates, Sunderland, Mass., pp. 340–61.

Brown, J.L. (1983) Cooperation – a biologist's dilemma. *Adv. in the Study of Behavior,* **13**, 1–37.

Brown, J.L. (1987) *Helping Behavior and Communal Breeding in Birds.* Princeton University Press, Princeton NJ.

Brown, J.L. and Brown, E.R. (1972) Helpers: effects of experimental removal on reproduction success. *Science,* **215**; 421–2.

Chagnon, N.A. (1974) *Studying the Yanomamo.* Holt, Rinehart and Winston, New York,

Chagnon, N.A. (1979) Mate competition, favoring close kin, and village fissioning among the Yanomamo Indians, In: *Evolutionary Biology and Human Social Behavior* (eds N. Chagnon and W. Irons), Duxbury Press, North Scituate, Mass., pp. 86–131.

Chagnon, N.A. (1982) Sociodemographic attributes of nepotism in tribal populations: man the rule breaker. In: *Current Problems in Sociobiology* (eds King's College Sociobiology Group). University of Cambridge Press, Cambridge.

Chisholm, J.S. (1981) Residence patterns and the environment of mother–infant interaction among the Navaho. In: *Culture and Early Interactions* (eds, T. Field, A. Sostek, P. Vietze and P.H. Leiderman), Lawrence Erlbaum Associates, Hillsdale, NJ, pp. 3–19.

Chisholm, J.S. (1983) *Navaho Infancy.* Aldine Press, New York.

Clarke, E. (1957) *My Mother Who Fathered Me.* Allen and Unwin, London.

Curry, R.L. (1987) Evolution and Ecology of Cooperative Breeding in Galapagos Mockingbirds, (Nesomimus *spp.*). Ph.D. Dissertation, University of Michigan, Ann Abor, Mich.

Daly, M. and Wilson, M.I. (1981) Abuse and neglect of children in evolutionary perspective. In: *Natural Selection and Social Behavior: Recent Research and New Theory* (eds R.D. Alexander and D.W. Tinkle), Chiron Press, New York. pp. 405–16.

Daly, M. and Wilson, M.I. (1983) *Sex, Evolution and Behavior,* 2nd edn, Willard Grant Press, Boston.

Daly, M. and Wilson M.I. (1985) Child abuse and other risks of not living with both parents. *Ethology and Sociobiology,* **6**, 197–210.

Daly, M. and Wilson, M.I. (1987) *Homicide,* Aldine Press, Hawthorne, New York.

Denham, W.W. (1978) BEVRECS. In: *Alyawara Ethnographic Data Base.* HRAF Press, New Haven.

Dickemann, M. (1981) Paternal confidence and Dowry competition: a biocultural analysis of purdah. In: *Natural Selection and Social Behavior: Recent Research and New Theory* (eds R.D. Alexander and D.W. Tinkle). Chiron Press, New York.

Draper, P. (1977) Social and economic constraints on child life among the !Kung. In: *The Kalahari Hunter-gatherers* (eds R.B. Lee and I. DeVore). Harvard University Press, Cambridge.

Draper, P. and Harpending, H. (1987) Parental investment and the child's environment. In: *Parenting Across the Lifespan: Biosocial Dimensions* (eds J.B. Lancaster, J. Altmann, A.S. Rossi and L.R. Sherrod), Aldine de Gruyter, New York, pp. 207–36.

Dublin, H.T. (1983) Cooperation and reproductive competition among female African elephants In: *Social Behavior of Female Vertebrates* (eds S. Wasser and M. Waterhouse), Academic Press, New York, pp. 291–313.

Emlen, S.T. (1982a) The evolution of helping. I. An ecological constraints model. *Am. Naturalist,* **119**, 29–39.

Emlen, S.T. (1982b) The evolution of helping. II. The role of behavioral conflict. *Am. Natur.* **119**, 40–53.

Emlen, S.T. (1984) Cooperative breeding in birds and mammals. In: *Behavioral Ecology,* 2nd edn (eds J.R. Krebs and N.B. Davies), Sinauer Associates, Sunderland, Mass., pp. 305–39.

Flinn, M.V. (1983) Resources, Mating, and Kinship: the Behavioral Ecology of a Trinidadian Village, Ph.D. thesis, Department of Anthropology, Northwestern University. University Microfilms, Ann Arbor, Mich.

Flinn. M.V. (1986) Correlates of reproductive success in a Caribbean village. *Hum. Ecol.,* **14**(2), 225–43.

Flinn, M.V. (1988a) Parent–offspring interactions in a Caribbean village: daughter guarding. In: *Human Reproductive Behaviour,* (eds L. Betzig, M. Borgerhoff Mulder and P. Turke). Cambridge University Press, Cambridge.

Flinn, M.V. (1988b) Mate guarding in a Caribbean village. *Ethnology and Sociobiology* **9**(1), 1–29.

Flinn, M.V. (1988) Step and genetic parent/offspring relationships in a Caribbean village. *Ethology and Sociobiology,* **9**(3), 335–69.

Flinn, M.V. and Low, B.S. (1986) Resource distribution, social competition and mating patterns in human societies. In: *Ecological Aspects of Social Evolution,* (eds D.I. Rubenstein and R.W. Wrangham), Princeton University Press, Princeton, NJ, pp. 217–43.

Gonzalez, N.S. (1969) *Black Carib Household Structure.* University of Washington Press, Seattle.

Gross, D. (1984) Time allocation: a tool for the study of cultural behavior. *A. Rev. Anthrop. 1984,* **13**, 519–58.

Hames, R. (1979) Relatedness and interaction among the Ye'kwana: a preliminary analysis. In: *Evolutionary Biology and Human Social Behavior* (eds N. Chagnon, and W. Irons), Duxbury Press, North Scituate, Mass., pp. 238–50.

Hames, R. (1988) The allocation of parental care among Ye'kwana. In *Human Reproductive Behaviour* (eds L. Betzig, M. Borgerhoff Mulder and P. Turke), Cambridge University Press, Cambridge.

Hamilton, W.D. (1964) The genetic evolution of social behavior I and II. *J. Theor. Biol.* **7**, 1–52.

References

Hrdy, S.B. (1976) Care and exploitation of non-human primate infants by conspecifics other than the mother. *Advances in the Study of Behavior*, **6**, 101–58.

Hrdy, S.B. (1981) *The Woman that Never Evolved*. Harvard University Press, Cambridge, Mass.

Irons, W. (1979) Investment and primary social dyads. In: *Evolutionary Biology and Human Social Behavior* (eds N. Chagnon and W. Irons), Duxbury Press, North Scituate, Mass., pp. 181–212.

Irons, W. (1983) Human female reproductive strategies. In: *Social Behavior of Female Vertebrates* (eds S. Wasser and M. Waterhouse), Academic Press, New York, pp. 169–213.

Issacs, G. (1978) The food sharing behavior of protohuman hominids. *Scient. Am.*, **20**, 90–108.

Jarvis, J.U.M. (1981) Eusociality in a mammal: cooperative breeding in naked mole-rat colonies. *Science*, **212**, 571–3.

Johnson, A.W. (1975) Time allocation in a Machiguenga community. *Ethnology*, **14**, (3), 310–12.

Johnson, A.W. (1978) *Quantification in Cultural Anthropology*. Stanford University Press, Stanford.

Keesing, R.M. (1976) *Cultural Anthropology*. Holt, Rinehart and Winston, New York.

Konner, M.J. (1975) Relations among infants and juveniles in comparative perspective. In: *The Origins of Behavior, Volume 3: Friendship and Peer Relations* (eds M. Lewis and L.A. Rosenblum), John Wiley and Sons, New York, pp. 99–129.

Lamb, M.E. and Sutton-Smith, B. (eds.) (1982) *Sibling Relationships: Their Nature and Significance Across the Lifespan*. Erlbaum, Hillsdale, NJ.

Lancaster, J.B. and Lancaster, C.S. (1987) The watershed: change in parental investment and family formation strategies in the course of human evolution. In: *Parenting Across the Lifespan: Biosocial Dimensions* (eds J.B. Lancaster, J. Altmann, A.S. Rossi, and L.R. Sherrod), Aldine de Gruyter, New York, pp. 187–206.

Lee, P.C. (1987) Allomothering among African elephants. *Animal Behavior*, **35**, 278–91.

Leiderman, P.H. and Leiderman, G.F. (1977) Economic change and infant care in an East African agricultural community. In: *Culture and Infancy: Variations in the Human Experience* (eds P.H. Leiderman, S.R. Tulkin and A. Rosenfield), Academic Press, New York, pp. 405–38.

LeVine, R. (1962) Witchcraft and co-wife proximity in southwestern Kenya. *Ethnology*, **1**, 39–45.

LeVine, R. (1977) Child rearing as cultural adaptation. In: *Culture and Infancy: Variations in the Human Experience* (eds P.H. Leiderman, S.R. Tulkin and A. Rosenfield), Academic Press, New York, pp. 15–27.

Lovejoy, C.O. (1981) The origin of man. *Science* **211**, 341–50.

Mayer, P. (1981) Biocultural Evolution of Human Longevity, PhD Dissertation, Department of Anthropology, University of Colorado, Boulder, Co.

McClintock, M.K. (1971) Menstrual synchrony and suppression. *Nature*, **229** 244–5.

McKenna, J.J. (1981) Primate infant-caregiving behavior: Origins, consequences and variability with emphasis upon the common Indian langur. In: *Parental Care in Mammals* (eds D. Gubernick and P. Klopfer), Plenum, New York.

McKenna, J.J. (1987) Parental supplements and surrogates among primates: Cross-species and cross-cultural comparisons. In: *Parenting Across the Lifespan: Biosocial Dimensions,* (eds J.B. Lancaster, J. Altmann, A.S. Rossi and L.R. Sherrod), Aldine de Gruyter, New York, pp. 143–84.

Moehlman, P.D. (1983) Socioecology of silverbacked and golden jackals. In: *Recent Advances in the Study of Mammalian Behavior* 7 (eds J.F. Eisenberg and D.G. Kleiman), Special Publication of the American Society of Mammalogy, pp. 423–53.

Moehlman, P.D. (1986) Ecology of cooperation in canids. In: *Ecological Aspects of Social Evolution* (eds D.I. Rubenstein and R.W. Wrangham), Princeton University Press, Princeton, NJ, pp. 64–86.

Munroe, R.H. and Munroe, R.I. (1971) Household density and infant care in an East African Society. *J. Soc. Psychol.,* **85**, 3–13.

Munroe, R.H. and Munroe, R.L. (1984) Household density and holding of infants in Samoa and Nepal. *J. Soc. Psychol.* **122**, pp. 135–6.

Munroe, R.L., Munroe, R.H., Michealson, C., Koel, A. Bolton, R. and Bolton, C. (1983) Time allocaton in four societies. *Ethnology,* **22**, 355–70.

Murdock, G.P. (1967) Ethnographic Atlas: A summary. *Ethnology,* **6**. 109–236.

Otterbein, K.F. (1966) *The Andros Islanders.* University of Kansas, Lawrence.

Parker, G.A. and McNair, M.R. (1979) Models of parent–offspring conflict. IV. Suppression: Evolutionary retaliation by the parent. *Animal Behavior,* **27**, 1210–35.

Quiatt, D. (1986) Juvenile/adolescent role functions in a rhesus monkey troop: an application of household analysis to non-human primate social organization. In: *Primate Ontogeny, Cognition, and Social Behavior* (eds J.G. Else and P.C. Lee), Cambridge University Press, Cambridge, pp. 281–90.

Reidman, M.L. (1982) The evolution of alloparental care and adoption in mammals and birds. *Q. Rev. Biol.,* **57**, 405–35.

Rood, J.P. (1978) Dwarf mongoose helpers at the den. *Zeitschrift für Tierpsychologie,* **48**, 277–8.

Rood J.P. (1986) Ecology and social evolution in the mongooses, In: *Ecological Aspects of Social Evolution* (eds D.I. Rubenstein and R.W. Wrangham), Princeton University Press, Princeton, NJ, pp. 131–52.

Shepard, T.H. and Fantel, A.G. (1979) Embryonic and early fetal loss. *Clin. Perinatology,* **6** (2), 219–43.

Skutch, A.F. (1987) *Helpers at Birds' Nests.* University of Iowa Press, Iowa City.

Smith, M.G. (1962) *West Indian Family Structure,* University of Washington Press, Seattle.

Stacey, P.B. (1979) Kinship, promiscuity, and communal breeding in the acorn woodpecker. *Behavioral Ecol. Sociobiol.,* **6**, 53–66.

Strassmann, B.I. (1986) M.S. Thesis. Cooperative Breeding in Humans and other Animals. Museum of Zoology, University of Michigan, Ann Arbor, Mich.

Super, C. (1981) Behavioral development in infancy. In: *Handbook of Cross-Cultural Human Development* (eds R.H. Munroe, R.L. Munroe and B.B. Whiting). Garland STPM Press, New York.

Trivers, R.L. (1974) Parent–offspring conflict. *Am. Zoologist,* **14**, 249–65.

Tronick, E.Z., Morelli, G.A. and Winn, S. (1987) Multiple caretaking of Efe (Pygmy) infants. *Am. Anthropologist,* **89**, 96–106.

Turke, P. (1985) Fertility determinants on Ifaluk and Yap: Tests of economic and Darwinian hypotheses. PhD Dissertation, Department of Anthropology, Northwestern University, University Microfilms, Ann Arbor, Michigan.

Turke, P. (1988) Helpers at the nest: childcare networks on Ifaluk. In: *Human Reproductive Behavior* (eds L. Betzig, M. Borgerhoff Mulder, P. Turke). Cambridge University Press, Cambridge.

Turke, P. and Betzig, L. (1986) Those who can do: wealth, status, and reproductive success on Ifaluk. *Ethology and Sociobiology,* **6**, 79–87.

Voland, E. (1984) Human sex-ratio manipulation: Historical data from a German Parish. *J. Hum. Evolution,* **13**, 99–107.

Wasser, S. and Barash, D.(1983) Reproductive suppression among female mammals: Implications for biomedicine and sexual selection theory. *Q. Rev. Biol.,* **58**, 513–38.

Weisner, T.S. (1982) Sibling interdependence and child caretaking: a cross cultural view. In: *Sibling Relations: Their Nature and Significance Across the Lifespan* (eds M. Lamb and B. Sutton-Smith), Lawrence Erlbaum Associates, Hillsdale, NJ, pp. 305–27.

Weisner, T.S. (1987) Socialization for parenthood in sibling caretaking societies. In: *Parenting Across the Lifespan: Biosocial Dimensions* (eds J.B. Lancaster, J. Altmann, A.S. Rossi and L.R. Sherrod), pp. 237–70. Aldine de Gruyter, Hawthorne, New York.

Weisner, T.S. and Gallimore, R. (1977) My brother's keeper: Child and sibling caretaking. *Current Anthropology,* **18**, 169–90.

West-Eberhard, M.J. (1983) Sexual selection, social competition, and speciation. *Q. Rev. Biol.,* **58**, 155–83.

Whiting, B.B. and Whiting, J.W.M. (1975) *Children of Six Cultures.* Harvard University Press, Cambridge Mass.

Williams, G.C. (1966) *Adaptation and Natural Selection.* Princeton University Press, Princeton.

Wilson, E.O. (1971) *The Insect Societies.* Harvard University, Belknap Press, Cambridge, Mass.

Wilson, E.O. (1975) *Sociobiology.* Harvard University, Belknap Press, Cambridge, Mass.

Wilson, M., Daly, M. and Weghorst, S. (1980) Household composition and the risk of child abuse and neglect. *J. Biosoc. Sci.,* **12**, 333–40.

Woolfenden, G.E. (1975) Florida scrub jay helpers at the nest. *Auk,* **92**, 1–15.

Woolfenden, G.E. (1981) Selfish behavior by Florida scrub jay helpers. In: *Natural Selection and Social Behavior: Recent Research and New Theory* (eds R.D. Alexander and D.W. Tinkle), Chiron Press, New York, pp. 247–60.

CHAPTER THIRTEEN

The sociocultural biology of Netsilingmiut female infanticide

Colin Irwin

13.1 INTRODUCTION

The Arctic environment is so actively hostile to life that the traditional Eskimo found they must not only adapt their technology, for hunting, clothing and shelter, to the needs of survival, but, it is suggested, they also had to tailor their social relationships and beliefs to the same needs (Rasmussen, 1931; Weyer, 1932; Balikci, 1968). In the traditional culture the most unfortunate of these adaptations was the killing of as much as 50% of all baby girls in tribes like the Netsilingmiut (Boas, 1901; Rasmussen, 1931). This high rate of female infanticide invites numerous moral, social and biological questions.

Sex ratio manipulation by means of infanticide is a well documented and much discussed topic in both sociobiology and cultural anthropology (Hausfater and Hrdy, 1984). Female infanticide has received attention both in studies of stratified societies (Dickeman, 1979) and in hunter gatherer societies (Birdsell, 1968; Chagnon, 1968; Dickeman, 1975; Chagnon *et al.*, 1979). Female infanticide amongst the Eskimo from Alaska to Greenland has been reviewed by Weyer (1932) and Schrire and Steiger (1974) with considerable discussion being focused on the Netsilingmiut of the Central Canadian Arctic by Boas (1901), Rasmussen (1931), Balikci (1967, 1968) and Freeman (1970, 1971). Birdsell (1972) provides a brief review and analysis of female infanticide amongst the Caribou Eskimo.

By combining previous studies with new data on female infanticide among the Netsilingmiut, an attempt is made here to demonstrate how both cultural anthropology and sociobiology are indispensable to a causal understanding of this phenomena. The evolutionary biologist's distinction between 'ultimate' causes (the selective advantages leading to the trait's prevalence), and 'proximate' causes (the environmental factors, events, structures and reflexes producing the adaptive response) is a useful one that warrants systematic extension to culturally produced behaviours. Questions about the proximate causes or mechanisms of behaviour can be characterized as being 'How?' questions while ultimate causes can be understood as 'Why?' questions (Barash 1982; Hausfater and Hrdy, 1984). The 'Why?' of Netsilingmiut female infanticide will be addressed first.

13.2 ULTIMATE CAUSE: ALTERNATIVE THEORIES

Weyer (1932), Balikci (1967, 1968) and Birdsell (1972) concluded that the practice of female infanticide is adaptive for the Eskimo, in culling the population, to maintain an ecological balance between the human predators and their Arctic prey. Following Williams' (1966) general critique most sociobiologists would now reject such a hypothesis as implying group selection (in the context of female infanticide, see Bates and Lees, 1979; Hawkes, 1981). Instead the current orthodoxy is to argue that all organisms, including humans, seek to maximize individual inclusive fitness (Bates and Lees, 1979), which may involve a relatively high parental investment in a small number of progeny as opposed to a relatively low parental investment in a comparatively large number of progeny. This evolutionary calculus may improve the fitness of a mother practising infanticide: Blurton Jones and Sibly (1978) analyse this behaviour amongst the !Kung; MacArthur and Wilson (1967) and Wilson (1975) give a theoretical analysis of general r and K selection theory). However Schrire and Steiger (1974) and Rasmussen earlier, have argued that female infanticide was maladaptive and was leading to population extinction in some cases. Since female infanticide was systematically practised in nearly all traditional Eskimo cultures, from Alaska to Greenland, and since the overall culture of the Eskimos may be considered a triumph of sociocultural adaptation to extreme circumstances, any simple dogma of maladaptive superstition seems unwarranted, and this chapter will seek out functional explanations. Again, female infanticide might have been an effort to produce an adaptive sex ratio. Biologists have attempted to apply theories of individual selection (Fisher, 1930) and kin selection (Trivers and Willard, 1973) to the problem of adjusting sex ratios in order to maximize reproductive success. It is argued here that no one of these theories taken simply is adequate to explain the infanticide data although all point in the correct direction.

If Eskimo female infanticide is a response to environmental pressures (Weyer, 1932: Balikci, 1967; Freeman, 1970, 1971; Birdsell, 1972) then rates of female infanticide should be correlated with indices of environmental severity. Further, in seeking the most appropriate application of adaptive sex ratio manipulation theory (section 13.3), these data will also be compared with rates of male mortality, adult sex ratios, marriage patterns and the division of labour by sex.

Female infanticide amongst the Eskimo has been reported by many anthropologists, ethnographers and explorers (Jackson, 1880; Nelson, 1899; Smith, 1902; Boas, 1907; Stefansson, 1914; Jenness, 1922; Rasmussen 1927; Birket-Smith, 1929; Weyer, 1932). It has even been reported as a rare contemporary phenomenon through female infant neglect (Williamson, 1974) referred to as 'deferred infanticide' by Johansson (1984). The ratio of girls to boys given in Table 13.1 implies that female infanticide was common,

Table 13.1
Eskimo juvenile and adult sex ratios

Location or tribe	No. of girls	No. of boys	Girls/ 100 boys	No. of women	No. of men	Women/ 100 men
1 Cape Prince of Wales (Weyer, 1932)	46	50	92	29	30	97
2 Cape Smyth (Smith, 1902)	14	27	52	52	45	116
3 Bernard Harbour (Jenness, 1922)	18	21	86	42	46	91
4 Netsilingmiut (Boas, 1907)	66	138	48	123	119	103
5 Sinamiut (Boas, 1907)	7	12	58	13	13	100
6 Sauniktumiut (Boas, 1907)	33	41	80	58	46	126
7 Qaernermiut (Birket-Smith, 1929)	11	24	46	30	25	120
8 Coast Padlimiut (Birket-Smith, 1929)	26	31	84	28	22	127
9 Interior Padlimiut (Birket-Smith, 1929)	20	28	71	31	25	124
10 Hauneqtormiut (Birket-Smith, 1929)	10	13	77	18	13	129
11 Harvaqtormiut (Birket-Smith, 1929)	15	23	65	21	17	123
12 Qaenermiut (Boas, 1907)	27	38	71	46	35	131
13 Avilikmiut (Boas, 1907)	15	27	56	34	26	131
14 N. Greenland exclusive of Thule (Birket-Smith, 1928)	773	803	96	2321	2918	115
15 South Greenland (Birket-Smith, 1928)	1106	1058	104	2801	2421	116
16 East Greenland (Birket-Smith, 1928)	117	118	99	234	211	111
17 East Greenland (Hansen, 1914)	128	99	129	175	146	120

although to varying degrees, across Arctic North America. These figures were compiled by Weyer (1932) from data collected by Smith (1902), Boas (1907), Hansen (1914), Jenness (1922), Birket-Smith (1928, 1929) and Weyer (1932).

The figures given in Table 13.1 also show how the situation of more boys than girls is turned around in the adult population to give more women than men. This reversal is attributed to a subsequent high mortality rate among young male hunters inexperienced in the techniques of Arctic survival (Nansen, 1893; Mauss and Beuchat, 1904–5; Bilby, 1923; Birket-Smith, 1928). This observation is further supported by additional data (Table 13.2) on the sex ratio of Eskimo populations in Greenland which show a female bias, irrespective of age, attributed to a combination of high male mortality and low rates of female infanticide (Weyer, 1932). These figures are compiled from population statistics collected by Birket-Smith (1928). Additionally Weyer tells us that:

'The death rate among young men is especially high; within the period dealt with by Bertelsen, proportionally four and a half times as many men of the age group thirty to thirty-five died as did within the same age-group in Denmark. This enormous rate of mortality among younger men is due primarily to the dangers of their hunting activities. Drowning in kayaks is

Table 13.2
Eskimo sex ratio in Greenland for total population in 1923 (Birket-Smith, 1928)

Location or tribe	No. of females	No. of males	Females/ 100 males
Upernivik	586	523	112
Umanak	731	682	107
Godhavn	188	170	111
Ritenbenk	296	284	104
Jacobshavn	324	271	120
Christianshaab	284	271	105
Egedesminde	807	772	105
Total North Greenland West Coast	3198	2973	108
Holstensborg	401	413	97
Sukkertoppen	696	580	120
Godthaab	644	612	105
Frederikshaab	493	417	118
Julianehaab	1837	1583	116
Total South Greenland West Coast	4071	3605	113
Total West Coast	7269	6578	111

the fate of many. In South Greenland in 1889 the death rate among males from this cause alone was 9.3 per thousand male population, or almost as high as the death rate from all causes in the United States in recent years.'

In the analysis to follow, I am going to take these data seriously, not as infallible but rather as the best we have on a very important phenomenon under extreme and rare circumstances. It should be noted that this approach is in marked contrast with the use which Schrire and Steiger make of these data. From Weyer's assemblage Schrire and Steiger find only Boas's observation of the Netsilik (data set no. 4) acceptable, rejecting the others as insufficiently well attested to for their purposes. They retain the data that is typical of the extremely cold, high infanticide cultures. Thus the presumed lower quality of the other similar cultures does not lead to any differences of interpretation. But they have excluded all of the warmer temperature, high female rate cultures that support the functional interpretations offered in this paper, and have inferred an unexplained maladaptive ethnocidal atavism. As Naroll (1962) has argued, poor data quality is usually a more plausible explanation of zero correlations than of significant relationships, unless one can come up with a specific hypothesis as to systematic biases in the poor quality of the data that would produce the relationship found. In reconsidering Weyer's sources, no such general explanations have suggested themselves (an exception for Greenland is noted later), nor have Schrire and Steiger offered any.

If male mortality was a contributing cause to the practice of Eskimo female infanticide this would in turn require an explanation. In traversing the North American Artic by dog team (Irwin, 1974) I became aware of the great heterogeneity of the environment with respect to severity of climate and associated abundance of game. This led me to the hypothesis that severity of environment might be related to differences in the rate of infanticide. Figure 13.1 maps the data points from Table 13.1. Two candidate features of the environment seem likely to be related to severity in producing a high death rate among young hunters. These are low temperatures and lack of light during much of the year. Given a finite capacity for foraging effort and efficiency the number of dependents an Eskimo hunter will be able to support will, to a considerable degree, be dependent on the abundance of game. The size, distribution and population density of caribou, birds and trout are dependent on plant life whose growth is restricted by low temperature and limited sunlight. The availability of sea mammals is also restrained by temperatures, as this affects sea ice cover which in turn limits access to whales and walrus which unlike seals do not make and maintain winter breathing holes.

Table 13.3 compiles the data used for the observation points given in Figure 13.1 along with annual mean temperature (US Department of Commerce 1965; Orvig, 1970), latitude (which is an index of available

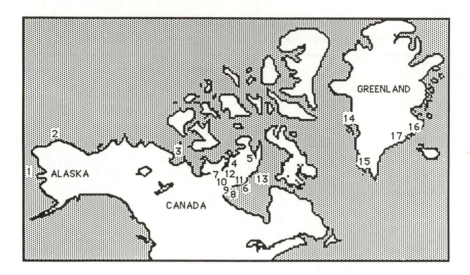

Figure 13.1 Map illustrating the distribution of data points for the index of female infanticide, and other variables, given in Table 13.3 and Table 13.3(b).

Table 13.3

Demographic and environmental aspects of Eskimo megapopulation: Location (Loc), Girls/100 boys(Girl), Women/100 males (Wom), Annual mean temperature (Temp), Male mortality index (Mort), Latitude (Lat)

Loc	Girl	Wom	Temp	Mort	Lat
1	92	97	−4.9	5	65.5
2	52	116	−12.5	64	71.5
3	86	91	−11.5	5	67.8
4	48	103	−13.7	55	69.5
5	58	100	−13.7	42	69.5
6	80	126	−11.3	46	63.4
7	46	120	−11.9	74	64.3
8	84	127	−9.1	43	61.0
9	71	124	−9.1	53	61.0
10	77	129	−11.9	52	64.3
11	65	123	−11.9	58	64.3
12	71	131	−11.9	60	64.3
13	56	131	−11.1	75	66.5
14	96	115	−3.9	19	70.0
15	104	116	+1.8	12	61.0
16	99	111	−0.4	12	66.0
17	129	120	−0.4	−9	66.0

Table 13.3(b)
Averaging of geographically grouped data points from Figure 13.1

Loc	Girl	Wom	Temp	Mort	Lat
1	92	97	−4.9	5	65.5
2	52	116	−12.5	64	71.5
3	86	91	−11.5	5	67.8
4	53	102	−13.7	49	69.5
(4 & 5 above)					
5	71	126	−11.0	55	63.2
(6,7,8,9,10,11 & 12 above)					
6	56	131	−11.1	75	66.5
7	107	116	−0.7	9	66.0
(14,15,16 & 17 above)					

sunlight), and a variety of derived indices. Let us first consider the correlations between the simpler variables and the environmental indicators as shown in Table 13.4. This table, and its reduced form Table 13.4(b) present the major finding central to the issue of this paper. The correlation of 0.85 between the ratio of girls/100 boys and mean temperature is focal. It is also presented graphically in Figure 13.2.

A potential difficulty with the data is Galton's problem (Naroll, 1970), created by the geographic grouping of the observation points. In an effort to minimize this difficulty the data from adjacent groups sharing the same culture have been averaged. The data for the Netsilingmiut subpopulations (observations 4 and 5) have been combined into one point. The group of points on the west coast of Hudson's Bay (6 to 12), and the Greenland observations (14 to 17) have similarly been pooled. The resulting seven data sets are given in Table 13.3(b) and the correlations in Table 13.4(b).

The most significant results are obtained for the correlation of juvenile sex ratio, and annual mean temperature giving a coefficient of 0.85 significant at the $p < 0.001$ level for the 17 observation point data set (Figure 13.2). This level of correlation is maintained at 0.88 with a significance of $p < 0.01$ for the seven observation point data set. The adult sex ratio also correlates with latitude for the 17 point data set at the $p < 0.04$ level of significance for a coefficient of −0.50. However this correlation is reduced to −0.27 for the seven point data set with no significance at $p < 0.554$.

There are no direct indicators of male mortality available in Weyer's, or others', population counts. In support of the male mortality argument the only quantitative data we have are the observation that whereas childhood sex

Table 13.4
Pearson's product moment correlations for data given in Tables 13.3 and 13.3(b).
Computed from Table 13.3 ($n = 17$)

Variables	Temperatures	Latitude
Girls/100 boys	+0.85***	−0.25
Women/100 men	−0.07	−0.50*
Male mortality index	−0.76***	−0.02

***$p < 0.001$
*$p < 0.05$
(Other correlations, Temp/lat, −0.25; Girls/women, −0.07. Some correlations involving the male mortality index are uninterpretable due to part–whole relationships and correlated error.)

Table 13.4(b)
Computed from Table 13.3(b) ($n = 7$)

Variables	Temperature	Latitude
Girls/100 boys	+0.88**	−0.51
Women/100 men	−0.03	−0.27
Male mortality index	−0.64	−0.23

**$p < 0.01$
(Other correlations, Temp/lat, −0.45; Girls/women, −0.034. Some correlations involving the male mortality index are uninterpretable due to part–whole relationships and correlated error.)

ratios range from 46 to 129 (Table 13.1), with only two showing a prevalence of girls (indices above 100), the adult sex ratios (Table 13.1) range from 91 to 131, with 15 of the 17 ratios showing a prevalence of women. In an exploratory analysis, this comparison was turned into a doubly derived index of male mortality although it should be noted that this index is subject to correlated error if correlated with either of the variables from which it is derived, juvenile and adult sex ratios. However correlated error is not a problem so long as the index of male mortality is only correlated with the geophysical and environmental variables latitude and temperature. Had male mortality been directly measured, and had the theory (section 13.3) been correct, that male mortality was a causally intermediate variable between temperature and female infanticide, it would be expected that the male mortality index would correlate more highly with temperature than did the ratio with girls. However, the correlation is slightly lower, 0.76 as opposed to 0.85 (0.64 in

241

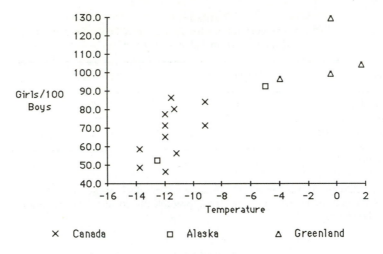

Figure 13.2 Annual mean temperature plotted against the index of female infanticide, ratio of girls/100 boys, for the data points given in Tables 13.1 and 13.3, and Figure 13.1.

contrast with 0.88 for the reduced, $n = 7$, analysis). Given the indirect nature of the index, the errorfulness of the components, this slight difference is not regarded as ruling out the hypothesized causal order. Note, for example, that for the sex ratios of the girls and women to be completely comparable, a cultural-ecological stability for the two generations would have to be assumed.

Low temperature, as an index of the harshness of the environment, would seem to be one of the contributing causes of female infanticide among the Eskimo. In contrast, the very low correlations with latitude are probably due to the warm ocean currents, that flow north around Alaska and Greenland, providing an ample supply of marine mammals. This fact would seem to more than compensate for the costs of darkness, making the brighter yet colder latitudes of the central continental Arctic significantly less hospitable to human life.

As the thesis of this chapter rests on the conclusions of this analysis it should be pointed out that these significant correlations do not stand alone but are supported by the ethnographic reports of many observers in different parts of the Artic at different times. All these reports attribute the sex and age biasing of the population statistics to female infanticide, with this necessitated by high male mortality. In the absence of more independent data, which is difficult or impossible to come by, the recorded accounts of cause of death provided by the researchers and their Eskimo informants can be taken into account in weighing the merits of this argument. There is here a method-ological triangulation between the statistical correlation and the reports of the explorers and anthropologists.

Schrire and Steiger (1974) point out that some Eskimo infanticide data, in particular those data collected more recently than that used here, will be influenced by prohibitions on infanticide introduced by western missionaries and legal agencies. This may be partly true of the data set used here as Weyer (1932) points out that the low rate of female infanticide in Greenland may be due to such influences. Even if these four data points are taken out of Table 13.3 a correlation of 0.68 is produced, which is significant at the $p < 0.01$ level for the remaining 13 observation points. Given the strength of the correlation, its stability under sample combination and truncation, the fact that the data was compiled by Weyer with no view to demonstrating a temperature correlation, and the breadth and depth of the supportive ethnographic record, two conclusions may be drawn as follows:

(1) Eskimo female infanticide is probably temperature dependent.
(2) Male mortality is the most likely agent in restoring balance to the adult Eskimo sex ratio.

13.3 ULTIMATE CAUSE: THEORETICAL DISCUSSION

Although the extreme severity of Arctic existence would seem to be a significant contributing cause of Eskimo female infanticide, this correlation does not explain why sex ratio manipulation is the appropriate response to the environmental pressure. For an explanation of this, evolutionary theories of sex ratio must be examined. Table 13.5 provides a summary of the theories and their implications (For a review from which some of the arguments presented here are taken, see Clutton-Brock and Albon, 1982.) Prominent amongst these theories is Fisher's (1930), which is based on the premise that the mean reproductive success of each sex is inversely related to their frequency in the population. From this thesis Fisher makes two conclusions that are pertinent to an analysis of the Eskimo data. (1) If the cost of producing male and female progeny are the same, the average parent will produce a 50:50 sex ratio. (2) If the cost of producing one sex exceeds the cost of producing the other sex, then the sex ratio will be biased against the more costly sex.

As Fisher pointed out, this second conclusion becomes more complex when mortality is sexually biased. In humans, the cost per male born will be less than the cost per female born if the males die more frequently before parental investment is completed. However the cost per male successfully raised to maturity will be more than it is for females as some investment in males will have been unproductive and must be added into the cost of rearing males. This could produce, as Fisher observed in humans, a male biased sex ratio at birth and a female biased sex ratio at maturity (Hypothesis 1, Table 13.5).

The Eskimo data would at first appear to be an extreme example of this phenomenon, inasmuch as the combined effects of female infanticide and

Table 13.5

Sociobiological theories of sex ratio manipulation and Arctic optimal foraging strategies

Hypothesis	Predictions	Results
(1) Approximately equal costs for producing males and females (Fisher, 1930)	Approximately 50/50 sex ratio at maturity	Yes. Slightly male biased birth sex ratio and female biased adult sex ratio
(2) Strongly biased costs for raising either sex (Fisher, 1930)	Far fewer of more costly sex will be raised	No. More costly males preferred and raised
(3) Sex ratio favours more altruistic sex (Hughes, 1981, from Trivers and Willard, 1973)	Male cooperators remain together. More of cooperating sex	Yes. Eskimo are patrilocal No. Fewer adult males
(4) Bias in sex ratio unaffected after parental care finished (Maynard Smith, 1978)	High male mortality produces polygamy	No. Most marriages are monogamous
(5) Phylogenetic inertia precludes nursing females from hunting (6) Cultural inertia precludes females from acquiring hunting skills	Only males hunt large mammals limiting the size of the nuclear family	Yes. In spite of male mortality most marriages are monogamous

male mortality does produce a slightly female biased adult population. However it can be argued that the cost per male born in the traditional Eskimo society exceeds the cost of each female born as the data suggest that half the female pregnancies are terminated at birth to make way for a male pregnancy (Rasmussen, 1931), with the result that it effectively takes one and a half pregnancies to produce a Netsilingmiut male infant (Hypothesis 2, Table 13.5). This issue is further complicated by what may be an arbitrary decision, on the part of biologists, as to the time of termination of parental care and hence the amount of parental investment (Clutton-Brock and Albon, 1982). The Netsilingmiut are patrilocal, so that mature but inexperienced males may continue to receive parental care after their sisters have left home. This bias in parental investment may be compensated for in part by the systematic adoption by the parents of their daughter's first son, so that the daughter's individual fitness is contributed to in spite of her absence.

Computing the costs and benefits of female pregnancy termination, patrilocal son investment and systematic daughter's progeny maybe so difficult as to make the application of Fisher's theory to this problem impossible.

However, Fisher's theory would seem to explain the approximately 50:50, male/female, birth sex ratio in humans, a ratio, which, for the Eskimo, would appear to be maladaptive.

Based on the paper by Trivers and Willard (1973) that postulates some of the circumstances under which a deviation from a 50:50 sex ratio may be of benefit to an organism, Hughes (1981) has produced a model for adaptive sex ratio manipulation that may be applied to hunter gatherer's such as the Eskimo (Dickeman (1979) gives an evolutionary model of human sex ratio manipulation in stratified societies). Essentially the Hughes model develops the idea that parents will manipulate the sex ratio so as to favour the more altruistic sex, the sex which contributes more to future generations through their cooperation. This model predicts that if the net benefits of brotherly cooperation (m_{max}) are compared with the net benefits of sisterly cooperation (f_{max}), with m_{max} greater than f_{max}, it will be to the benefit of the parent to practise female infanticide. As Hughes correctly points out this situation may frequently arise in patrilocal societies that practise cooperative hunting such as the Netsilik and Yanomamo (Hypothesis 3, Table 13.5). Unfortunately the Eskimo cooperators die at such an alarming rate that the adult population has a slight female bias. This fact does not necessarily disprove Hughes' model, but it does suggest that Hughes thesis is not a complete explanation in this particular case.

Leigh (1970) and Maynard Smith (1978) point out that if sex differences in mortality continue after the termination of parental investment then the optimum sex ratio will not be affected inasmuch as the diminished group of survivors will be individually more successful the fewer they are. In a human context this thesis would predict that polygynous relationships should develop in societies with high male mortality (Hypothesis 4, Table 13.5). The high frequency of polygyny in numerous human situations adds weight to this thesis but it is not a common solution among the Eskimo for dealing with their sex ratio problem. For example, of 58 Netsilingmiut families Rasmussen (1931) surveyed in the Central Arctic in 1923, 54 were monogamous, one was polyandrous and only three were polygynous. The Netsilingmiut do not have cultural prohibitions on adult males or females from having more than one partner so why do not adult males, that are surviving beyond parental care, regularly have more than one wife, as a cultural alternative to female infanticide? Perhaps the answer to this question lies in the unusual degree of division of labour, by sex, to be found among the Eskimo?

In general hunting, trapping, and catching large aquatic animals is an almost exclusively male activity across cultures (Murdock, 1937; Murdock and Provost, 1973; Ember, 1981). This is explained to be due to universal constraints on women's productive activity, due to childbirth and nursing (Ember and Ember, 1971; Burton *et al.*, 1977; White *et al.*, 1977; Ember, 1981; White *et al.*, 1981). These observations and conclusions suggest two biological restraints, one phylogenetic, the other ecological, that may contribute to the Eskimo division of labour by sex (Hypothesis 5, Table 13.5). (1) Human sexual dimorphism restricts the nursing of infants to females. (2) Arctic foraging, which is frequently limited to large mammals, generally excludes nursing females from hunting activities.

These two restraints could possibly produce a situation in which nearly all the food gathering is done by the males. In general biological terms this may be understood as facultative parental behaviour which frequently results in monogamy when successful reproduction requires the cooperation of two committed adults. This is particularly true of carnivores as the males of these species can carry food of high calorific value back to the young and it is doubly true of species, like Antarctic penguins, that must forage while taking care of their offspring in a hostile environment (Barash, 1982). For the Netsilingmiut the number of wives and progeny the male is able to support is dependent on his foraging ability and in the Arctic the dangers of hunting are such that it appears to take, on average, the sequential lives of two Netsiling-miut males to support one female and her progeny. In this kind of model a parent's fitness can be seen to be maximized by constraints similar, but not identical to, those operating in Fisher's theory. If an excess of females are produced they will lack males to provide for them and they and their progeny will die. If insufficient females are produced then parents who raise more daughters will be favoured by natural selection.

The restraining effect nursing has, on the interchangeability of sex rolls, may be a sufficient explanation for the sexual division of labour in this case. If so further argument will be redundant; however, Schrire and Steiger (1974) conclude:

'Although the practice is no longer current, we are bound to suggest a far more efficient way of culling these populations, by simply modifying the traditional sexual division of labor. Instead of insisting that only men hunt, women might have been trained for the task. This type of role exchange did occur traditionally under extraordinary conditions (Kemp, pers. comm.) but normally only males were regarded as potential pro-viders.'

It could be argued that females are not morphologically so well equipped, as males, to the rigours of hunting, and it should be noted that such activities can produce miscarriages (Irwin, 1985). Additionally both male and female Eskimos would probably find Schrire and Steiger's suggestion for cultural modification unworkable as they would be quick to point out that women don't know how to hunt and if a woman died in a hunting accident the man would be unable to look after her orphaned infants. Childcare in the Arctic is far from easy:

'It was about that time when I almost lost my little daughter. She almost froze. I must have been so clumsy. I had gone home to sleep with my husband. That is when I almost froze my little daughter. I woke my husband up and we cried when no one else could hear us. My husband died that summer . . .

When it was about two years old my father told me to give it to my cousin because he knew that I could not look after the baby well. I did not know how to take care of a baby. My cousin had many babies, but they all died when they were a few days or a year old, she never had any success with a healthy baby. That is why my father told me to give the baby to her, because I could not look after it as well.'

(Aupudluck, 1977)

These observations highlight what may be yet another restraint on the exchange of sex roles, namely, cultural specialization (Hypothesis 6, Table 13.5). In the Arctic the technologies of hunting, clothing manufacture and childcare are so sophisticated no one individual is able to become proficient at all these skills. Thus not only does an Eskimo require a mate for reproduction but Eskimo may find it difficult to even survive without the food or clothing provided by a member of the opposite sex. As learning these separate skills begins in infancy and as their teaching is retained within each sex, a gradual evolution toward some females acquiring hunting technology could not take place. In other words, it may be speculated that the possibility of teaching females to hunt was prohibited by cultural inertia, in much the same way as phylogenetic inertia kept males from acquiring the capacity to nurse infants. The evolution of either capacity, hunting females or nursing males, would have required radical and substantial changes in the Eskimo coadaptive culturtype and/or genotype. This, it is suggested, did not happen because moving from one adaptive peak to another, in both cultural and/or genetic terms, frequently requires traversing a maladaptive valley (for genetic topographic models of fitness, upon which this argument for cultural inertia is based, see Wright (1929, 1970, 1977). Such a model, when applied to epistemic evolution, may equally well describe the dynamics of paradigm shift (Kuhn, 1962) in the sociology of knowledge. The comparison, however, is not a perfect analogy, as cultural change can sometimes leap a maladaptive valley by, for example, the process of cultural contact and change, acculturation).

Other ultimate causes could possibly be added to the ones mentioned here that are suggested as being instrumental in forging the phenomenon of Eskimo female infanticide. Hopefully the principal elements are those identified in Table 13.5. Although causes associated with Hypotheses 2, 3 and 4 may effect the Eskimo population they are probably not as important as the causes associated with Hypotheses 1, 5 and 6 in producing the very high rates of female infanticide to be found. However this analysis concerning the why of

Eskimo female infanticide can only be part of a possible explanation of the phenomena. The question of how the behaviour occurs must now be examined.

13.4 PROXIMATE CAUSES

Questions concerning the 'how' of Eskimo female infanticide are questions that centre on the proximate causes or mechanisms of behaviour, in contrast with the ultimate evolutionary causes. Mechanisms for altering progeny sex ratios in invertebrates are common (Hamilton, 1967; Trivers and Hare, 1976; Charnov *et al.*, 1981), but problems of phylogenetic inertia with regard to the physiology of sex determination may prevent the evolution of such mechanisms in mammals (Beatty, 1970; Williams, 1979; Maynard Smith, 1980). Some examples of possible adaptive mammalian sex ratio manipulation, including human, have been suggested, but they are generally thought not to be significant (Clutton-Brock and Albon, 1982). The possibility that male sex ratio bias in the Eskimo may be in part a genetic trait that can be manipulated physiologically before birth cannot be totally ruled out. The human sex ratio is slightly male biased and this tendency can possibly be moved a few percentage points under some environmental conditions (Parkes, 1926; Teitelbaum, 1972; McMillen, 1979). However the Netsilingmiut would seem to require a 2 to 1 male biased birth sex ratio and perhaps this extreme imbalance could not evolve due to phylogenetic limitations. Additionally ethnographic descriptions of the Eskimo attribute the biased infanticide sex ratio to female infanticide. This begs the question as to what the proximate mechanisms or causes of this behaviour are and how did they evolve? It is conceivable that this behaviour was largely genetically controlled but the speed with which modern Eskimos abandoned female infanticide, when confronted with the novel environment created by modern western culture, makes this hypothesis doubtful. The proximate causes are therefore very likely cultural in which case the beliefs associated with the behaviour of female infanticide would be the likely proximate mechanisms. It should be pointed out here that culture is not limited to being a mere instrument of behaviour; it is much else besides, for example, culture is a repository for knowledge, and a vehicle and mechanism for the refinement of that knowledge. However examining the processes of cultural evolution with regard to vehicles, mechanisms and units of selection, for the culture traits to be described here, is beyond the scope of this chapter.

The methods of cognitive and cultural anthropology can be used to identify and describe beliefs. Each belief can be considered to be a culture trait, and those traits considered to be pertinent to the behaviour under examination here could be termed a coadaptive culture type for female infanticide. Several such culture traits have been identified: a value, a rationalization, a metaphysic, and a cognate. These elements of Netsilingmiut, and frequently Eskimo culture in general, are to be found in many Arctic ethnographies (Irwin, 1981 and 1985, gives the ethnographic methodology).

In order to facilitate the association of certain values or ethical notions of praiseworthy or blameworthy behaviour with certain acts it is necessary to provide cognitive separations of such acts. As in English, the Eskimo language distinguishes the taking of human life into the praiseworthy or accepted categories, such as execution, and unacceptable or immoral categories, such as murder. Eskimo differs from English in classifying suicide as acceptable, and in having no cognate for war. Directly relevant to the problem of this paper is the fact that infanticide is acceptable, providing the infant has not been named, and has a separate term, contrasting with murder. Table 13.6 gives these terms and cognates.

The Eskimo metaphysics of reincarnation produce a situation in which the qualities of personhood are synonymous with the name. (Weyer's (1932) review encompasses the whole Arctic, Irwin (1981) gives an analysis of this metaphysic with respect to Eskimo ethics; see also Balikci (1967).) In traditional times names, and with them sanctity and personhood, were normally given by the oldest, most senior relatives. When a child was born and until it was named, the grandparents generally had the authority of life and death over the baby. Providing no extraordinary circumstances such as starvation prevailed, the infant would be promptly named if it was a boy. However, it would only receive a name if it was a girl under the most favourable conditions, such as plenty of food, a promise of future marriage, or a general surfeit of boys in that particular family group. In more recent times, now that female infanticide is no longer a required practice, naming is done during pregnancy so that sanctity and personhood is given to the fetus before the sex is known. (The names employed are traditionally given without regard to the sex of the baby or the ancestor previously holding the name.)

In spite of the appearance of a certain stoic cruelty in these matters, it should be understood that the event of female infanticide was always accompanied with considerable emotional difficulty. Such deaths were perceived as necessary, 'they could not be helped,' but the accompanying distress, particularly for the mother, was not perceived as in any sense good. To take one example from my own family-by-marriage: Aupudluck had thirteen children of which three are now living. The grandparents would not give her first daughter by her third and present husband a name. The parents wanted to keep the child but would not go openly against the wishes of the older relatives and therefore could not give the baby a name themselves. Consequently the infant existed without sanctity and personhood for several months, until an unrelated elderly friend took pity on her and gave the infant a name. Unfortunately, that girl died when she was six. But because she was loved so much her name was given to the next daughter when she was born. Aupudluck's husband, Kako, explained, 'Infanticide was mostly for the baby girls

Table 13.6

Netsilingmiut cognates for the taking of human life

English	Eskimo and literal translation	
Baby abandonment	Nutaraarluk* Baby	Iksingnaoktauyoq Leave it – he or she – doing it
Baby freezing	Nutaraarluk Baby	Qiqititauyoq Freeze it – he or she – doing it
Suicide	Inminik Self	Pitariok He or she took life
Murder	Inuaktok Human – he or she took life	
Revenge killing or execution	Akeyauok Back – he or she – to him or her	
Two people fight	Unatuktook Fight – them (two)	
Many people fight	Unatuktoon Fight – them (many)	
War**	None	
Many people fighting** and killing each other	Unatuktoon Fight – them (many) – human – they took life (many)	Inuakgrotioon

* A baby that is not a person can be discriminated by the qualification that the baby does not have a name/soul, Nutaraarluk Atikungitok: Baby – Name – he or she– without.
** With a holoplastic language, likè Inuktitut, it is possible to generate almost any meaning. However 'War' as an organized activity that does not entail the blameworthy taking of life (murder) can not be given an adequate translation. Revenge killing or execution, Akeyauok, can only be used with respect to specific acts of murder and the associated murders.

when there were too many girls and not enough boys. Those infants didn't have names. Sometimes they were named before birth but naming during delivery was most popular. Nowadays the names are often given during pregnancy.'

Balikci (1967), reaches essentially the same conclusions as those presented here, i.e. the practice is an adaptation to the harsh environment. However, he also concludes as does Freeman (1970, 1971), that it is to some degree a sexist act, inasmuch as females are less valued than males. While in my judgement this is not totally true for adult Eskimos or girls after they are named, it is undoubtedly true for unnamed infants. Still, today, the birth of a boy is seen as an occasion for much praise of the mother, while the birth of a girl is frequently treated with relative indifference.

In addition to the metaphysical manipulation of the sanctity and the sexually biased value of infants, the rational analysis of their predicament may also influence the occurrence of Eskimo female infanticide. Rasmussen's experience as an ethnographer and explorer, who lived off the land in the Arctic, provides a unique insight to the logic employed.

'A hunter must take into consideration that he can only subject himself and his constitution for comparatively few years to all the strain that hunting demands. Competition is keen, and if he has no very special natural gifts and enjoys no unusually good health, he need not be very old before he can no longer hold his own with the young. Now if he has sons, they will as a rule be able to step in and help just when his own physique is beginning to fail. Thus it is life's own inexorability that has taught them the necessity of having as many sons as possible. Only by that means may they be certain that they will not need to put the rope round their own neck too early; for it is the common custom that old people, who can no longer keep themselves, prefer to put an end to their life by hanging rather than drag themselves through life in poverty and helpless old age.

Nalangiaq once said to me: "Life is short. We all want to be as prosperous as we can in the time we are alive. Therefore parents often consider that they cannot 'afford' to waste several years nursing a girl. We get old so quickly, and so we must be quick and get a son. That is what we parents think, and in the same way we think for our children. If my daughter Quertilik had a girl child I would strangle it at once. If I did not, I think I would be a bad mother.' (Rasmussen, 1931)

Nalangiaq's analysis does not master the maths of Fisher's theory or similar evolutionary stable strategies that may model the dynamics of sex ratios so as to maximize fitness in future generations (MacArthur, 1965). Nonetheless the Eskimo's logical conclusions concerning the need for female infanticide possibly illustrate an ability to approach a sensitive moral issue empirically. However the quotation from Rasmussen's informant, Nalangiaq, does raise two more issues that may require comment. Firstly, although it has been argued that Eskimo sex ratio manipulation promotes individual fitness, including the mother's, more often than not it is the grandparents who order the infanticide. This aspect of the behaviour may be in part due to a need for a dominant member of the extended family to override the natural maternal instincts of the mother. However, a single mother who kept all her daughters, while those around her killed some of theirs, could be at an advantage. This possibility may be circumvented by a mother-in-law making sure a daughter-in-law gains no reproductive benefits over her own children in whom she may have a greater genetic interest. Thus, unlike Fisher's individual fitness model, the analysis made here may also have to be viewed from the perspective of the individual and/or inclusive fitness of the grandparents whose interests can be different to those of the mother (Trivers, 1974; Alexander, 1974). Secondly, although Nalangiaq and Rasmussen rationalize the benefits of female infanticide in terms of practical advantages, the perceived function is not increased fitness in future generations. This observation does not necessarily undermine the thesis developed here as the content of any belief can be of less importance than the belief's adaptive value, in evolutionary terms. However, if Eskimo female infanticide was practised, as Nalangiaq and Rasmussen suggest, so as to systematically eliminate the first females born to any family, then the first children to mature in any family would be male hunters and providers. In this way the manipulation of the age, as well as sex structure, of the Eskimo population, may also be adaptive as pointed out by Freeman (1971).

Some sociobiologists might still wish to argue that the culture traits for female infanticide identified here could be genetically determined insofar as they could be part of a genetically preferred coadaptive culture type (Lumsden and Wilson, 1981). Two observations would seem to cast doubt on the application of such a hypothesis in this instance. First, and presumably maladaptively, the mother of the infant that is put to death is distressed, sometimes to the point of rebellion against those ordering the infanticide. Second, now that the Eskimo have been brought under the care of a welfare state, female infanticide is almost non-existent, and the associated culture traits have been modified or abandoned.

It should be noted, that in traditional times, it can be suggested that there always was a biological evolutionary selective pressure for a male biased sex ratio among the Eskimo. Parents that produced more males would be more successful, as they would have fewer female pregnancies wasted through the culturally instituted practice of female infanticide. It might be thought that, in the perhaps 300 generations that they have lived in the Arctic, this might have led to an innate bias in the Eskimo birth sex ratio. On the other hand, an analysis is possible in which cultural proximate mechanisms are distinctly superior to genetic determinates. Note that the Eskimos do not live in a homogeneous environment, as indicated by the data in Table 13.3 and Figures 13.1 and 13.2. Nor is the Eskimo megapopulation static, inasmuch as it has been created by successive waves of migration from Siberia and Alaska through Canada to Greenland (Giddings, 1967). These migrations have taken place approximately every millennium. Thus the Eskimo population have been continually moving through an annual mean temperature gradient of some 15°C, requiring different optimal rates of female infanticide, according to the adaptive interpretation made here. The presence of large amounts of ancient whale bone in the Central Arctic, where the Netsilingmiut presently live, suggest that the people of that region used to hunt whales. The severity of summer ice cover precludes the penetration of many species of whales into these regions today. This suggests that the Central Arctic climate was warmer in its not too distant past, and that the inhabitants could have possibly maximized their reproductive success with a lower rate of female infanticide then. These observations on climatic changes and migrations of the megapopulation would place the Eskimo in an environment in which optimal sex ratio fluctuated by decades or centuries. These time frames may be too short, in evolutionary terms, for a genetic proximate mechanism to adjust to the changing optimal sex ratio appropriate to changing environmental conditions, with sufficient speed (Lumsden and Wilson's '1000 year rule', 1981.) In this circumstance a cultural proximate mechanism, with its inherent plasticity and fine tuned by its rational components, has considerable adaptive advantage over a genetic mechanism (Campbell, 1965; Mason, 1979; Burhoe, 1981; Pulliam and Dunford, 1980; Shields, 1982; Boyd and Richerson, 1983). Unfortunately this cultural process remains maladaptive in terms of the superfluous female pregnancies, but then both phylogenetic and cultural inertia are such that no natural solution to the pressures of selection will ever be perfect.

13.5 TRACING THE EVOLUTIONARY AND RECENT NATURAL HISTORY OF NETSILINGMIUT FEMALE INFANTICIDE

In an effort to rationalize schematically the sequence of ultimate and proximate causes that lead to the behaviour of Netsilingmiut female infanticide I found I inevitably ended up with a model that was always wrong at some point, Figure 13.3a, or always correct but trivial, Figure 13.3b. Tracing the causal chain of any complex human behaviour is probably destined to failure as the chain is in reality a shifting causal network. Nonetheless a model of some sort, within which to systematize the analysis made here, would be welcome, in order to understand more clearly what has just been done and provide a framework within which to build similar analysis of other human behaviours.

Tracing the historical origins of phenomena can frequently help us focus on the most relevant causes of phenomena. Cosmology is dominated by inquiries into the history of the universe. Likewise tracing the history of ideology, and documenting human sociocultural history, provides us with important insights into contemporary cultures and social practices as they are manifest. In biology the historical approach, to the analysis of adaptive function, has been advocated by Pittenbrigh (1958), Williams (1966), Mayr (1974), and Barash (1982). This analytic method may be of help in developing a better understanding of human behaviours in general (Tinbergen 1968; Irwin, 1985, 1986).

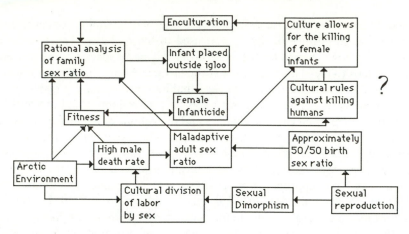

Figure 13.3(a) 'Causal network' for the behaviour of Eskimo female infanticide? Probably incorrect.

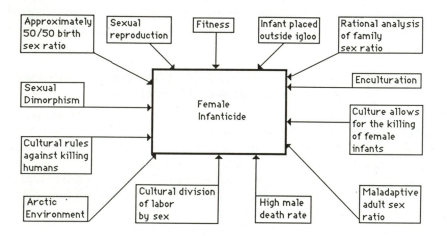

Figure 13.3(b) The causes of Eskimo female infanticide? Probably correct but ambiguous and trivial.

Figure 13.4 takes the causes of Netsilingmiut female infanticide, identified in this chapter, and attempts to arrange them in order of their evolutionary emergence and recent natural history. By way of a thought experiment I will speculatively review this history in an effort to illustrate some of the potential benefits of the synthesis being attempted here. Firstly, and perhaps always, we must begin with the imperatives of life and fitness, most commonly understood in terms of reproductive success and survival. Relevant to the history of the behaviour being examined here the advent of sexual reproduction is an

Figure 13.4 'Evolutionary spectrum'

Category bands (vertical labels across the spectrum):
FITNESS | SEXUAL DIMORPHISM | ENVIRONMENT | ADULT SEX RATIO / ADAPTIVE RADIATION | ENCULTURATION | ADAPTIVE RADIATION | FEMALE INFANTICIDE

FITNESS		ENVIRONMENT		ENCULTURATION		FEMALE INFANTICIDE
Aprox 50/50 birth sex ratio	Cultural rules against killing humans	Cultural division of labor by sex	High male death rate	Culture allows for the killing of female infants	Rational analysis of family sex ratio	Infant placed outside igloo

GENOTYPE <--> PHENOTYPE

GENETIC EVOLUTION <--->CULTURAL EVOLUTION <---> INDIVIDUAL DEVELOPMENT <--> PHENOTYPE /BEHAVIOR

PHYLOGENY <----------------------------> ONTOGENY <----------------------------> PHENOTYPE /BEHAVIOR

ULTIMATE CAUSE <---------------------- PROXIMATE CAUSE <---------------------- PHENOTYPE /BEHAVIOR

SOCIOBIOLOGY <--------> CULTURAL ECOLOGY <--------> COGNITIVE, DEVELOPMENTAL, SOCIAL ANTHRO/PSYC.

Figure 13.4 'Evolutionary spectrum' for the sociocultural biological analysis of Eskimo female infanticide. The model is generated by tracing the events, relevant to the behaviour being examined, through their evolutionary and recent natural history.

important event. In our mammalian past our birth sex ratio was set at approximately 50/50, male/female, and in our more recent past a number of factors (e.g. male competition for females, differences in male and female parental investment, extended nursing) contributed to human sexual dimorphism. It is difficult to know exactly when cultural controls restricting human egoism were first introduced into human societies. This may have occurred before hunter gatherers extended their range into the Artic, some seven to fifteen thousand years ago, as living in the Arctic required considerable cultural invention. As these migrants moved into increasingly less hospitable regions the techniques of hunting large Arctic mammals was developed and practised predominantly by males while females specialized in the manufacture of protective clothing. The rate of male mortality rose and more male hunters were needed to feed the females and their infants. Phylogenetic inertia prevented any physiological adjustment to the birth sex ratio while motherly instincts for progeny and cultural rules against the taking of human life prevented the adjustment of the birth sex ratio by female infanticide. Through some process of cultural evolution unnamed infants were defined as not being fully human, girls were valued increasingly less than boys and freezing or abandoning unnamed infants was not defined as murder with its usual penalty of execution at the hands of the deceased's male relatives. These culture traits were acquired in the youth of each Eskimo during the enculturation process. When these individuals matured to become the elderly dominant members of their extended families they cognitively analysed the needs of their family with respect to the desirability of keeping newborn male and female infants. This analysis was made within the framework of the previously acquired culture. On some occasions a female baby would not be named and would be placed outside the igloo to freeze. The behaviour and associated event of female infanticide was thus precipitated.

This historical abstraction of the major arguments presented in this chapter, generates a description of the behaviour that passes from genotype to environment to phenotype; from sociobiology through the various social sciences (e.g. cultural ecology, sociocultural and cognitive anthropology, developmental and social psychology) to behaviour, from genetic evolution through cultural evolution and individual development to behaviour, from phylogeny through ontogeny to behaviour and from ultimate cause through proximate cause to behaviour. This model may be of value in the identification of potentially important gaps in the description and explanation of any behaviour. It may also be of value in analysing the processes of environment and cultural change that have now resulted in the cessation of systematic female infanticide amongst the Eskimo. This question is very complex however, involving the introduction of the rifle, the fur trade, Christian ideology, legal sanctions and welfare. An analysis of the relative importance of these factors, in bringing about change, is beyond the scope of this chapter.

13.6 GENERAL DISCUSSION

The evolutionary theories of Fisher, Trivers and Willard, Maynard Smith and Hughes that seek to explain the dynamics of the phenomena of sex ratio would seem to be indispensable to providing a causal explanation of Eskimo female infanticide. However none of the theories appears to provide a complete analysis of the ultimate causes in this case as they fail to identify the importance of sexually biased Eskimo foraging. This of course is not the fault of these theoreticians as they were not writing specifically on Netsilingmiut female infanticide. However this fact helps emphasize the complexity of biological and cultural processes. Any complete sociobiological analysis may have to be species specific, site specific and time specific. This problem of specificity characterizes human sociobiology since humans live in such a wide range of environments in contrast with other animals.

Another major and possibly more serious difficulty with applying sociobiological theory to humans is the frequent failure to address the proximate causes of behavior. This criticism might be fairly levelled at Hughes' analysis of the evolution of sex ratio manipulation in hunter gatherer societies. Perhaps Hughes only wished to address the question of ultimate cause, but if so his presentation would have benefited by him saying as much. The proximate causes of human behaviour may frequently be cultural, due to phylogenetic restraints and the superior fine tuning that can only be achieved by the more rapidly responsive cultural mechanisms. Human sociobiologists frequently ignore the instrumental role culture plays in the occurrence of human behaviour. Ultimate causes may be an indispensable element in a scientific theory or analysis of any human phenomena. However sociobiologists should not be surprised when cultural anthropologists characterize their explanations as trivial, in the absence of specifying the role of cultural proximate mechanisms.

The analysis of Eskimo female infanticide by Weyer (1932) and more particularly the analysis of the behaviour amongst the Netsilingmiut by Balikci (1967, 1968) and among the Caribou Eskimo by Birdsell (1972) supports the thesis that the behaviour is due to environmental pressures. However these pressures are identified as the need for population control so that the Eskimo population would not exceed the carrying capacity of their habitat. Although it is widely accepted that reproductive strategies can balance the selective advantage of a high investment in a small number of progeny against a small investment in a large number of progeny, expressed in terms of 'r and K selection' (MacArthur and Wilson, 1967) most evolutionary biologists now accept the thesis that all organisms continually maximize their fitness and would not sacrifice their own progeny to benefit other members of the population (Williams, 1966; Wilson, 1975; Bates and Lees, 1979). But does it matter if cultural anthropologists do not always identify the precisely correct set of ultimate causes as long as they describe the cultural facts accurately? Is acquiring proficiency in a new and complex discipline like sociobiology to be considered essential?

These questions might be addressed from the philosophy of social science as to whether a certain theory is or is not scientific because it is or is not complete nomologically (Nagel, 1979; Rosenberg, 1980). This chapter seeks to resolve such issues by example rather than philosophic inquiry. The test of the relative merits of various analyses is made in terms which allows for a more or less useful interpretation of the facts. The analysis of Weyer and Balikci fails to address the complex dependent relationships that exist among birth sex ratio, sex role specialization, male mortality, environmental conditions and optimal foraging strategies. These are such, that the analysis made here would predict, that a change in any one of the variables should produce a change in the rate of female infanticide and subsequent changes in the beliefs associated with the pertinent cultural proximate mechanisms. Their lack of biological sophistication is no doubt due to the fact that their work preceded the recent growth of sociobiology. Even Birdsell (1972) and Schrire and Steiger (1974) are subject to this same difficulty. However, contemporary social science that fails to understand the ultimate causes of human behaviour is omitting an indispenable element from the explanation, prediction or control of the processes of cultural change. In this context cultural anthropologists should not be surprised if sociobiologists, in return, sometimes, perceive or characterize such work as being superficial.

For example, it has been pointed out to me (Rosenberg, 1985; Barkow, 1986) that a cultural materialist, such as Harris (1979), might accept the explanation given here that depends on the relationship between the Eskimo, their environment and means of production. Essentially that portion of the argument that includes, and falls to the right of, 'Arctic environment' in Figure 13.4. In response I would wish to argue that the cultural materialist explanation is substantially correct but incomplete, as the reason why the wasteful behaviour of female infanticide is adopted, with all its associated cultural proximate mechanisms, is because portions of the human biogram (Tiger and Fox, 1971), in this case birth sex ratio and sexual dimorphism, coupled with the biological restraints of phylogenetic inertia, prevent a more economic, less painful, physiological, adaptive manipulation, of the Eskimo birth sex ratio.

13.7 CONCLUSION

Human sociobiologists appear to primarily direct their attention to the biological 'why?' questions, while cultural anthropologists appear, by way of contrast, to be more concerned with cultural 'how?' or even descriptive 'what?' questions. This compartmentalization of disciplines results in incomplete or confused explanations of human phenomena. For example, in a critique of human sociobiology, the evolutionary theorist Gould (1976) fails to realize that sociobiological inclusive fitness interests are being served when elderly Eskimo commit suicide in order to release resources to their children and grandchildren, whether or not the mechanism controlling that behaviour is for the most part genetic or for the most part cultural. The perspective used here regards nature versus nurture to be a false dichotomy. Nature and nurture, biology and culture are interdependent phenomena (Fox, 1975; Katz *et al.*, 1974; Durham, 1976, 1979; Alexander, 1979; Bateson, 1982; Barash, 1982). Unfortunately the sciences of sociobiology and cultural anthropology neglect this fact in their respective descriptive titles. As a constructive step toward an integration of these perspectives one might better name the scientific study of man *sociocultural biology*. At the very least the kind of analysis of Netsilingmiut female infanticide made here can best be so characterized.

Such theory is not offered as a replacement for other social science or biological theories but rather as an extension of each of them, incorporating them into an evolutionary framework of cause and effect. In this way *co-adaptive culturetype* includes some of the explanatory principles of structural functionalism and cognitive anthropology. The individual and inclusive fitness interests of sociobiology may frequently translate into material and economic costs and benefits. However these translations would not be made for their own sake but for the purpose of providing a causal explanation. In this context it now seems reasonable to suggest that the kinds of explanations provided here for some Eskimo metaphysical beliefs, cognates, values and rationalizations may, in principle, be achieved for all the major beliefs, cognates, values and rationalizations of any culture (Campbell, 1975). Such an explanation does not necessarily demean these phenomena. Metaphysical beliefs that may be just plain wrong for a correspondence epistemology may nonetheless mirror a social system ecological necessity. They may frequently contain profound insights into the relation of humans to their human and physical environment in which the beliefs have evolved.

ACKNOWLEDGEMENTS

Firstly I must thank my Netsilingmiut wife Kunga, my mother-in-law Aupud-luck and my father-in-law Kako for providing me with a cultural understanding of an extremely sensitive issue. The Social Science and Humanities Research Council of Canada, Killam Trust at Dalhousie and New York State Board of Regents Schweitzer Chair at Syracuse are to be thanked for supporting my research with fellowships and travel costs. I must extend my gratitude to the scholars in philosophy, biology, anthropology, religious studies and linguistics who freely provided me with the knowledge to pursue this interdisciplinary study. These are principally, Donald Campbell, William Shields,

Alexander Rosenberg, Klaus Klostermaier, Michael Stack, Joan Townsend, William Starmer and Larry Wolf. The numerous anonymous scholars who made valuable suggestions on earlier drafts of this chapter, and undoubtedly saved me much embarrassment, must also be thanked with special gratitude going to Donald Campbell for his careful editing of the final manuscript.

REFERENCES

Alexander, R.D. (1974) The evolution of social behavior. *Ann. Rev. Syst. Ecol.*, **4**; 325–83.

Alexander, R.D. (1979) *Darwinism and Human Affairs.* University of Washington Press, Seattle.

Balikci, A. (1967) Female infanticide on the arctic coast, *Man*, **2**; 615–25.

Balikci, A. (1968) The Netsilik Eskimo: Adaptive Processes. In: *Man the Hunter* (ed. R.B. Lee and I. DeVore), Aldine, Chicago.

Barash, D.P. (1982) *Sociobiology and Behavior.* 2nd edn, Hodder and Stoughton, London.

Barkow, J. (1986) Personal communication at public lecture. Department of Sociology and Social Anthropology, Dalhousie University, January 17th.

Bates, D.G. and Lees, S.H. (1979) The myth of population regulation. In: *Evolutionary Biology and Human Social Behavior: An Anthropological Perspective* (eds N.A. Chagnon and W.G. Irons), Duxbury Press, North Scituate, Mass., pp. 86–131.

Bateson, P.P.G. (1982) Behavioral development and evolutionary processes. In: *Current Problems in Sociobiology* (ed. King's College Sociobiology Group), Cambridge University Press, Cambridge.

Beatty, R.A. (1970) Genetic basis for the determination of sex. *Philosophical Transactions of the Royal Society London*, **B259**, 3–13.

Bilby, J.W. (1923) *Among Unknown Eskimos: Twelve Years in Baffin Island.* S. Service, London.

Birdsell, J.B. (1968) Some predictions for the Pleistocene based on equilibrium systems among recent hunter-gatherers. In: *Man the Hunter* (ed. R.B. Lee and I. DeVore), Aldine, Chicago.

Birdsell, J.B. (1972) *An Introduction to the New Physical Anthropology.* Rand McNally and Co., Chicago.

Birket-Smith, K. (1928) The Greenlanders of the present day. In *Greenland*, a compilation of various authors, **II**, pp. 1–207.

Birket-Smith, K. (1929) *The Caribou Eskimos: Report of the Fifth Thule Expedition, 1921–1924.* Gyldendalske Boghandel, Copenhagen.

Blurton Jones, N.G. and Sibly, R. (1978) Testing adaptiveness of culturally determined behaviour: do bushmen/women maximize their reproductive success by spacing births widely and foraging seldom? In *Human Behaviour and Adaptation* (eds V. Reynolds and N. Blurton Jones), Symposium No. 18, Society for Study of Human Biology, Taylor and Francis, London, pp. 135–57.

Boas, F. (1901) The Eskimo of Baffin Land and Hudson Bay. *Bulletin of the American Museum of Natural History.* **XV**.

Boas, F. (1907) The Central Eskimo. *Bureau of American Ethnology, Annual Report No.*

6 (for 1884–1885) pp. 399–699.

Boyd, R. and Richerson, J. (1983) Why is culture adaptive? *Q. Rev. Biol.*, **58**, 209–14.

Burhoe, R.W. (1981) *Toward a Scientific Theology.* Christian Journals Ltd., Belfast.

Burton, M.L., Brudner, L.A. and White, D.R. (1977) A model of the sexual division of labor. *Am. Ethnologist,* **4**, 227–51.

Campbell, D.T. (1965) Variation and selective retention in socio-cultural evolution. In: *Social Change in Developing Areas: A Reinterpretation of Evolutionary Theory* (eds. H.R. Barringer, G.I. Blankstern and R.W. Mack), Schenkman, Cambridge, Mass.

Campbell, D.T. (1975) On the conflicts between biological and social evolution and between psychology and moral tradition. *Am. Psychol.*, **30**, 1103–26.

Chagnon, N.A. (1968) *Yanomamo: The Fierce People.* Holt, Rinehart and Winston, New York.

Chagnon. N.A., Flinn, M.V. and Melancon, T.F. (1979) Sex-ratio variation among the Yanomamo Indians. In: *Evolutionary Biology and Human Social Behavior: An Anthropological Perspective* (eds N.A. Chagnon and W.G. Irons), Duxbury Press, North Scituate, Mass., pp. 290–320.

Charnov, E.L., Hartogh, R.L., Jones, W.T. and van den Assem, J. (1981), Sex ratio evolution in a variable environment. *Nature,* **289**, 27–33.

Clutton-Brock, T.H. and Albon, S.D. (1982) Parental investment in male and female offspring in mammals. In *Current Problems in Sociobiology.* (ed. King's College Sociobiology Group), Cambridge University Press, Cambridge.

Devine, M. (1982) Outcrop Ltd, Editor, *N. W.T. Data Book: 1982–1983.* The Northern Publishers, Yellowknife, Canada.

Dickeman, M. (1975) Demographic consequence of infanticide in man. *Annual Review of Ecology and Systematics,* **6**, 107–37.

Dickeman, M. (1979) The reproductive structure of stratified human societies: A preliminary model. In: *Evolutionary Biology and Human Social Behavior: An Anthropological Perspective* (eds, N.A. Chagnon and W. G. Irons), Duxbury Press, North Scituate, Mass., pp. 321–67.

Durham, W.H. (1976) The adaptive significance of cultural behavior. *Hum. Ecol.*, **4**, (2), 89–121.

Durham, W.H. (1979) Toward a coevolutionary theory of human biology and culture. In: *Evolutionary Biology and Human Social Behavior: An Anthropological Perspective* (eds N.A. Chagnon and W.G. Irons), Duxbury Press. North Scituate, Mass., pp. 39–59.

Ember, C.R. (1981) A cross-cultural perspective on sex differences. In: *Handbook of Cross-Cultural Human Development* (eds R.H. Munroe, R.L., Munroe and B.B. Whiting), Garland Press, New York.

Ember, M. and Ember, C.R. (1971) The conditions favoring matrilocal versus patrilocal residence. *Am. Anthrop.*, **73**, 571–94.

Fisher, R.A. (1930) *The Genetical Theory of Natural Selection.* Oxford University Press. Oxford.

Fox, R.A. (1975) Primate and human kinship. In: *Biosocial Anthropology* (ed. R. Fox) John Wiley and Sons, New York.

Freeman, M.M.R. (1970) Ethos, economics and prestige: a re-examination of Netsilik Eskimo infanticide. *Proceedings of the 38th International Congress of Americanists,* **2**, 247–50.

Freeman, M.M.R. (1971) A social and ecologic analysis of sysematic female infanticide among the Netsilik Eskimo. *Am. Anthrop.*, **73**, 1011–18.

Giddings, J.L. (1967) *Ancient Men of the Arctic.* Knopf, New York.

Gould, S.J. (1976) Biological potential vs. biological determinism. *Natural History Magazine*, The American Museum of Natural History (May).

Hamilton, W.D. (1967) Extraordinary sex ratios. *Science*, **156**, 477–88.

Hansen, J. (1914) List of inhabitants of the East Coast of Greenland. *Meddelelser om Grønland*, **XXXIX**, pp. 181–202.

Harris, M. (1979) *Cultural Materialism.* Random House, New York.

Hausfater, G. and Hrdy, S.B. (1984) (eds) *Infanticide: Comparative and Evolutionary Perspectives.* Aldine, New York.

Hawkes, K. (1981) A third expansion for female infanticide. *Human Ecology*, **9** (1).

Hughes, A.L. (1981) Female infanticide: sex ratio manipulation in humans. *Ethnology and Sociobiology*, **2**, 109–11.

Irwin, C.J. (1974) Trek across Arctic America. *National Geographic*, **March**.

Irwin, C.J. (1981) Eskimo Ethics and the Priority of the Future Generation. Master's Thesis, Interdisciplinary Program, University of Manitoba.

Irwin, C.J. (1985) Sociocultural Biology: Studies in the Evolution of Some Netsilingmiut and Other Sociocultural Behaviors. Doctoral Dissertation, Department of Social Science, Syracuse University.

Irwin, C.J. (1986) A study in the evolution of ethnocentrism. In: *The Sociobiology of Ethnocentrism: Evolutionary Dimensions of Xenophobia, Discrimination, Racism and Nationalism* (eds V. Reynolds, V. Falger and I. Vine), Croom Helm, London.

Jackson, S. (1880) *Alaska, and Missions on the North Pacific Coast.* Dodd, Mead, New York.

Jenness, D. (1922) *Life of the Copper Eskimos*, Report of the Canadian Arctic Expedition, 1913–1918, XII.

Johansson, S.R. (1984) Deferred infanticide: excess female mortality during childhood. In: *Infanticide: Comparative and Evolutionary Perspectives* (eds G. Hausfater and S.B. Hrdy), Aldine, New York.

Kuhn, T.S. (1962) *The Structure of Scientific Revolutions.* University of Chicago Press, Chicago.

Leigh, E.G. (1970) Sex ratio and differential mortality between the sexes. *Am. Natur.*, **104**, pp. 103–10.

Lumsden, C.J. and Wilson, E.O. (1981) *Genes, Mind and Culture, The Coevolutionary Process.* Harvard University Press, Cambridge.

MacArthur, R.H. (1965) Ecological consequences of natural selection. In: *Theoretical and Mathematical Biology* (ed. T. Waterman and H. Morowitz), Blaisdell, New York, pp. 388–97.

MacArthur, R.H. and Wilson, E.O. (1967) *The Theory of Island Biogeography.* Princeton University Press, Princeton, N.J.

Mason, W. (1979) Ontogony of social behavior. In: *Handbook of Behavioral Neurobiology, Vol. 3. Social Behavior and Communication* (eds P. Marler and J.G. Vandenbergh), Plenum Press, New York.

Mauss, M. and Beuchat, M.H. (1904–5). Essai sur variations saisonnières des sociétés Eskimos, Étude de morphologie sociale, *l'Année Sociologique*, 9me.

References

Maynard Smith, J. (1978) *The Evolution of Sex.* Cambridge University Press, Cambridge.

Maynard Smith, J. (1980) A new theory of sexual investment. *Behav. Ecol. Sociobiol.,* **7**, 247–51.

Mayr, E. (1974) Teleological and teleonomic: a new analysis. *Boston Studies in the Philosophy of Science,* **14**, 19–117.

McMillen, M.M. (1979) Differential mortality by sex in fetal and neonatal deaths. *Science,* **204**, 89–91.

Murdock, G.P. (1937) Comparative data on the division of labor by sex. *Social Forces,* **15**, 551–3.

Murdock, G.P. and Provost, C. (1973) Factors in the sex division of labor. *Ethnology,* **12**, 203–25.

Nagel, E. (1979) *The Structure of Science.* Hackett Publishing Co., Cambridge.

Nansen, Fr. (1893) *Eskimo Life.* Longmans, Green, London.

Naroll, R. (1962) *Data Quality Control.* Free Press, New York.

Naroll, R. (1970) Galton's Problem. In: *Handbook of Method in Cultural Anthropology.* (eds R. Naroll and R. Cohen), Natural History Press, Garden City, New York.

Nelson, E.W. (1899) The Eskimo about Bering Strait. *Bureau of American Ethnology, 18th Annual Report,* Washington.

Orvig, S. (1970) *Climates of the Polar Regions: World Survey of Climatology.* Vol. 14, Elsevier, London.

Parkes, A.S. (1926) The mammalian sex ratio. *Biol. Revs,* **2**, 1–51.

Pittenbrigh, C.S. (1958) Adaption, natural selection and behaviour. In *Behaviour and Ecology* (eds A. Roe and G.G. Simpson), Yale University Press, New Haven, Conn.

Pulliam, H.R. and Dunford, C. (1980) *Programmed to Learn.* Columbia University Press, New York.

Rasmussen, K. (1927) *Across Arctic America.* Putnam, London.

Rasmussen, K. (1931) *The Netsilik Eskimos: Social Life and Spiritual Culture. Report of the Fifth Thule Expedition, 1921–24, Vol. VIII, No 1–2.* Gyldendalske Boghandel, Copenhagen.

Rosenberg, A. (1980) *Sociobiology and the Preemption of Social Science.* Johns Hopkins University Press, Baltimore, Maryland.

Rosenberg, A. (1985) Personal communication, dissertation defence. Dept. of Social Science, Syracuse University. April 25th.

Schrire, C. and Steiger, W.L. (1974) A matter of life and death: an investigation into the practice of female infanticide in the Arctic. *Man,* **9**, 161–84.

Shields, W.M. (1982) *Philopatry. Inbreeding and the Evolution of Sex,* State University of New York Press, Albany, N.Y.

Smith, M. (1902) In *The White World* (ed. R. Kersting), Lewis Scribner, New York, pp. 113–30.

Stefansson, V. (1914) The Stefansson–Anderson Arctic Expedition of the American Museum: Preliminary Ethnological Report. *Anthropological Papers of the American Museum of Natural History,* **XIV**, pt. 1.

Teitelbaum, M.S. (1972) Factors associated with the sex ratio in human populations. In: *The Structure of Human Populations* (eds G.A. Harrison and A.J. Boyce), Clarendon Press, Oxford.

Tiger, L. and Fox, R. (1971) *The Imperial Animal.* Holt, Rinehart and Winston, New York.

Tinbergen, N. (1968) On War and Peace in animals and man. *Science,* **160**. A.A.A.S. June 28th.

Trivers, R.L. (1974) Parent–offspring conflict. *American Zoologist,* **14** (1), 249–64.

Trivers, R.L. and Hare, H. (1976) Haploidiploidy and the evolution of the social insects. *Science,* **191**, 249–63.

Trivers, R.L. and Willard, D.E. (1973) Natural selection of parental ability to vary the sex ratio. *Science,* **179**, 90–2.

U.S. Department of Commerce (1965) *World Weather Records. 1951–60. Vol. 1 North America.* Washington, D.C.

Werner, O. and Campbell, D.T. (1970) Translating, working through interpreters, and the problem of decentering. In: *A Handbook of Method in Cultural Anthropology* (eds R. Naroll and R. Cohen), Natural History Press/Doubleday, New York.

Weyer, E.M. (1932) *The Eskimos: Their Environment and Folkways.* Yale University Press, New Haven.

White, D.R., Burton, M.L. and Brudner, L.A. (1977) Entailment theory and method: a cross-cultural analysis of the sexual division of labor. *Behav. Sci. Res.,* **12**, 1–24.

White, D.R., Burton, M.L. and Dow, M.M. (1981) Sexual division of labor in African agriculture: a network autocorrelation analysis. *Am. Anthropologist,* **83** (4), 824–49.

Williams, G.C. (1966) *Adaptation and Natural Selection.* Princeton University Press, Princeton, N.J.

Williams, G.C. (1979) The question of adaptive sex ratio in outcrossed vertebrates. *Proceedings of the Royal Society, Edinburgh,* **55**, 1–22.

Williamson, R.G. (1974) *Eskimo Underground: Socio-Cultural Change in the Canadian Central Arctic* Institution för Allnän och Jämförande Etnografi vid Uppsala Universitat, Uppsala.

Wilson, E.O. (1975) *Sociobiology: The New Synthesis.* Harvard University Press, Cambridge, Mass.

Wright, S. (1929) Evolution in a Mendelian population. *Arat. Record,* **44**, 287.

Wright, S. (1970) Random drift and the shifting balance theory of evolution. In: *Mathematical Topics in Population Genetics* (ed K. Kojima) Springer-Verlag, Berlin.

Wright, S. (1977) *Evolution and the Genetics of Populations.* A Treatise in Four Volumes. Vol. 3. University of Chicago Press, Chicago.

CHAPTER 14

The costs of children and the adaptive scheduling of births: towards a sociobiological perspective on demography

Nicholas G. Blurton Jones

14.1 INTRODUCTION

The cost of children has often been discussed by anthropologists and demographers as an important limiting factor for human reproduction (Nag *et al.*, 1978). The costs are usually considered as economic costs, detracting from the wealth of the family. Various other principles are invoked to explain why anyone has any children at all. The costs are seen as reasons to reduce natural, pre-set physiological fertility, or as reasons behind a wish or cultural pressure to attain a target family size. Sometimes they are seen as counteracting the wish to have sufficient children to provide for one's own economic well being in old age (a most unsociobiological purpose).

A sociobiologist starts by expecting maximization of the number of descendants. His problem is then the reverse, why might people have fewer than the physiological maximum number of children? The insights of parent–offspring conflict theory, with its suggestion that resources allocated to one child may detract from the ability to produce and raise others, suggests a way to relate costs of children to reproductive success, and to try to reconcile the two extreme positions. In this chapter I summarize an earlier attempt at such a combined examination of human reproduction – a study of the adaptiveness of birth spacing in the !Kung. I then discuss issues that arise when we begin to apply the same approach in another population, and put forward some simple models that might suggest other circumstances in which we might expect fitness maximization to restrain fertility.

The !Kung bushmen of northwest Botswana are noted for their long inter-birth intervals (IBI), averaging about four years between births (Lee, 1972; Howell, 1979), without the use of modern contraceptives. There has been much research on the physiological mechanisms by which this is achieved (Konner and Worthman, 1981; Howie and McNeilly, 1982). It seems to be generally agreed that very frequent suckling delays the return of ovulation after birth, and that this would be a major factor in the physiological control of bushman birth spacing. There seems to be growing evidence that probability of ovulation is also influenced by aspects of a woman's nutritional level, energy budget or exercise (Dobbing, 1985; Ellison *et al.*, 1986).

The question of whether such long IBIs are adaptive or not has been addressed by Blurton Jones and Sibly (1978) and Blurton Jones (1986, 1987). The questions were: are the long inter-birth intervals (IBI) of the !Kung evidence, as some suppose, that people do not maximize reproductive success? Or on the contrary, do the long intervals come about because they do indeed increase reproductive success? Does reducing the rate of production of offspring lead to such an increase in survivorship that it ultimately results in an increase in the number of offspring raised to maturity or at least in the rate of production of surviving offspring? First Blurton Jones and Sibly constructed a model of some of the possible costs of short IBIs. Then, assuming these costs translate into mortality of children, they proposed an optimal IBI, and predicted some other aspects of birth schedules. Tests of the predictions were reported by Blurton Jones (1986, 1987), using Howell's (1979) data.

14.2 ADAPTIVENESS OF !KUNG INTER-BIRTH INTERVALS

14.2.1 The Blurton Jones and Sibly backload model

During the late dry season in the Dobe area, !Kung women forage over long distances, predominantly visiting mongongo nut groves that may be as far as six miles from their dry season camp which is near the only available surface water. Later in the dry season, families move to the nut groves, obtaining water from roots for a few days then returning to the camp near a water hole for a while. Before this change in strategy, women cover great distances in a quite severe climate. Women carry small babies with them all the time, and frequently take children with them, who up to the age of about four, need to be carried for substantial portions of the journey. Sometimes children are left at camp, sometimes there are women on the foraging trip who can help carry the child for part of the time. The late dry season (September to November), when distance between water and nut groves is greatest, seems likely to be the hardest time of the year.

Lee (1972) showed how short intervals between births would result in a

!Kung women having to carry great weights of baby for many, many miles each year. If she had the next baby before the previous one was either walking or staying at home, she would be burdened with carrying two children for much of her time when out in the bush foraging, or when moving camp. Lee suggested that long IBIs avoided such intolerable levels of work.

Blurton Jones and Sibly repeated Lee's calculations, and added to them the food requirements of mother and children – using more figures reported by Lee. We produced two graphs (Figure 14.1), one of average weight of baby carried on a foraging journey against age of baby, the other of weight of the food (mongongo nuts) brought home by mother to contribute at the observed level (58.8% of the calories) to the feeding of that child up to the age of 15 years old. We then wrote a computer program that superimposes these graphs on each other with variable time lags – corresponding to IBIs, and sums the weight the mother will have to carry home from foraging trips at each point in her reproductive career to support the family that her schedule of inter-birth intervals produces (Figure 14.3). A woman who has a baby every two years could rapidly end up trying to support ten or more children at once, and trying to carry impossibly large loads of food and children. A woman who only has a baby every five years, on the other hand, will just produce her third when the first becomes independent at age 15, and will have much less to carry at all times. Fifteen may be an underestimate for the age of independence of !Kung children. Teenage children are apparently fed by their

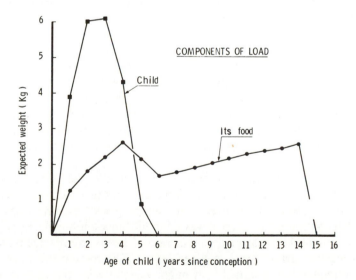

Figure 14.1 The weight carried by a mother for each child. Average weight of child carried shown separately from average weight of food carried to feed it.

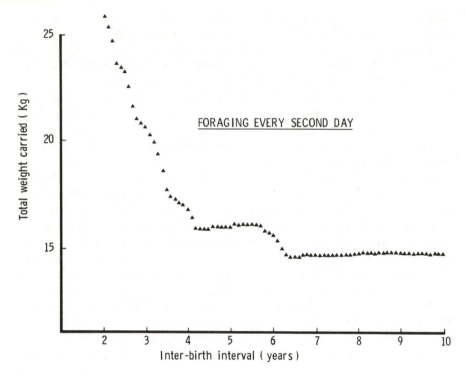

Figure 14.2 Calculated total weight of children and food for her family, carried by a mother if she collects food every second day.

parents for much longer than this (Draper, 1976). However, older children do a little baby sitting and teenage girls sometimes accompany their mother on gathering trips, so this underestimate may not distort the picture too badly.

These calculations confirm Lee's suggestion, and show the situation to be at least as difficult as Lee describes. The highest load carried by the mother at some point in her reproductive career increases very sharply for IBIs below four years (Figure 14.2).

Having shown how expected backload increases rapidly when IBIs below four years are adopted, and having discussed ways in which increased backload is costly, and even life-threatening, we proposed that this might indeed mean that the four-year interval was optimal. Since the weight to be carried increases massively for a small gain in rate of births, we may suppose these diminishing returns sooner or later reach zero. The costs of the much greater backload required to feed the family might well outweigh the gains in number of births. We went on to outline some of the factors to be taken into account in a direct test of this proposition.

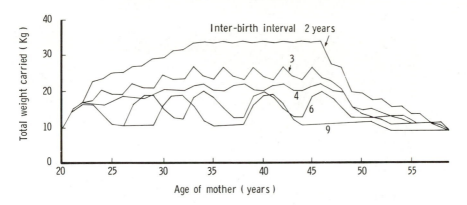

Figure 14.3 Graph of weight that the Blurton Jones and Sibly model calculates that a woman would carry if she maintained interbirth intervals shown (and kept other factors constant – described in Blurton Jones and Sibly, 1978), plotted against a range of possible IBIs.

14.2.2 A test for optimal IBI

In 1982 Nancy Howell agreed to let me use her reproductive histories of !Kung women to attempt such a test (Blurton Jones, 1986). The test was restricted to the sample to which the backload model should apply, i.e. !Kung women whose reproductive lives were spent living predominantly in the bush. Since the backload model also predicts (section 14.2.3) that first intervals will be under less constraint than later intervals, only later intervals were used. 'Replacement' intervals (when the first child dies before the next pregnancy) are also omitted from this analysis, because it seems clear that in these intervals mortality determines the intervals and not vice versa.

What curve of mortality against IBI was represented by the observations? Intervals were classed as 'success' if they added a surviving child to the family, and 'failure' if they did not (if either or both the children defining the interval died). Survival of a child to ten years old was chosen as a criterion for successful reproduction because subsequent mortality is very low. The raw data on success and failure was fitted (using the logistic regression program BMDPLR from the BMDP series, Dixon, 1981) to a curve described by IBI, 1/IBI, and to Backload calculated by the Blurton Jones and Sibly model for each IBI. Backload and 1/IBI gave the best fits according to several statistical tests. Thus mortality follows a curve that climbs more steeply as IBI decreases (Figure 14.4), and closely parallels the way backload increases as IBI decreases.

From these curves, the yield of surviving teenagers given by each IBI was calculated as follows: the number of births resulting from employing a parti-

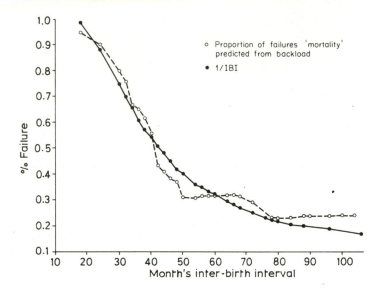

Figure 14.4 Fitted curves of failure ('mortality') predicted by backload and by 1/IBI, shown plotted against IBI, for ungrouped data. Results obtained by logistic regression.

cular IBI throughout a 20 year reproductive career was multiplied by the probability of 'success' of such intervals. This shows that an interval of 46–50 months would yield the greatest number of surviving teenagers, and is thus regarded as the optimal interval.

If natural selection has designed a reproductive system that is efficient, (and in particular responds efficiently, if mercilessly, to the mother's workload as a measure of potential mortality) one would expect that the observed inter-birth intervals would be normally distributed about the optimum interval. Figure 14.5 shows the curve of yield superimposed on a bar chart of the number of intervals of each length. Intervals of 48 months were indeed the most common interval for bush living women, a close match to the optimum of 46–50 months (Figure 14.5). Shorter intervals were the next most abundant. If bush living !Kung women fail to optimize, they fail more often by having intervals that are too short. The long intervals are not maladaptive.

14.2.3 Other predictions and tests from the backload model

Many more predictions were made by assuming that !Kung women and their reproductive systems would maximize the number of descendents. I assumed that higher backloads led to higher mortality but that times when backloads would be much lower than usual provided opportunities to shorten IBI and

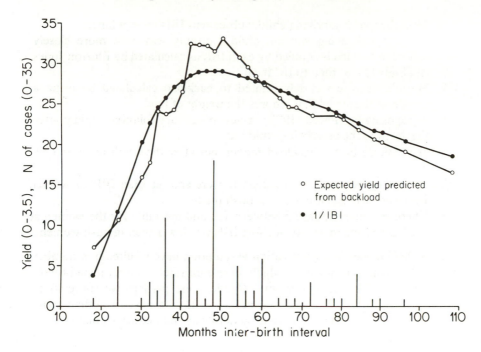

Figure 14.5 Graph of yield of surviving teenagers predicted by the best predictors (backload and 1/IBI), and bar chart of frequency of occurrence of each interbirth interval. Ungrouped data.

increase the number of births (as long as this did not bring backload up above usual levels). The predictions are not affected by the level chosen. One can conveniently think of 15–16 kg as a load commonly required, and frequently reported to be carried but any arbitrary level gives the following predictions. The predictions derive from two principles: that births should be maximized, but that high backloads must be avoided. Thus first intervals entail a much lower load, shortening them packs more births into the time and brings back-load up toward the 'limit'. So I expected that the first IBI would be shorter than later IBIs.

The predictions and their tests are reported in Blurton Jones (1987). As predicted by the model:

1. First inter-birth intervals (IBI) were shorter than were later IBIs.
2. The survivorship of children in first IBIs was not significantly related to length of IBI – shorter IBIs gave as good a survivorship as longer IBIs.
3. For subsequent IBIs, mortality increased markedly as IBI decreased.
4. As the family grows from the first surviving child to four surviving children, IBIs become longer.

271

5. After the fourth surviving child, subsequent IBIs are not longer.

6. For the bush living women, child mortality was even more closely related to backloads entailed by each IBI, as calculated by Blurton Jones and Sibly (1978) than to IBI itself.

7. Mortality was less closely related to backload calculated by using a version of the model that ignores the weight of food.

8. As reported by Howell (1979), a new pregnancy followed rapidly after the death of the preceeding child but

9. as predicted by the backload model, not after the death of older children.

10. IBIs of women living at cattleposts were shorter than IBIs of women who were dependent mainly on bush foods.

11. There was no relationship between IBI and mortality for the women at cattleposts, and mortality at short IBIs was lower than in bush women.

The backload model and optimization assumption have resulted in a number of testable predictions, many of which were confirmed. Mother's load therefore appears to have an important influence, and perhaps we successfully identified a major cost of child-rearing. I would like next to outline some criticisms of the work that relate to its generalization to other populations.

14.2.4 Criticisms of the !Kung study

Among the several possible criticisms of the above work, the following relate to the application of this approach to human populations in general:

(a) Harpending and Pennington (in press) have shown that inter-birth interval is not the only component of !Kung reproductive success and that total number of births or length or reproductive career explains more of the low fertility of the !Kung than do long IBIs. As far as I can see, their analysis does not disprove the association between IBI and mortality or the optimality of long IBIs. Howell's data (used here) show the same contribution of length of reproductive career to reproductive success as Harpending and Pennington describe (Blurton Jones, 1986).

(b) Perhaps the high mortality observed at short IBIs is merely a result of the age of the first child at the birth of the second (since in all populations younger children are more vulnerable than older ones). This is apparently not the explanation for the findings, for in the !Kung data the child that ends the interval shows the same high mortality at short intervals and low mortality at long intervals as reported above.

(c) If individuals in a population have different circumstances, their optimal strategies may differ and looking at the population as a whole may show no optimum. The population shows an optimum in the !Kung case but it might not have done, for instance if we had combined cattle post and bush food women. Evidently it is important to identify important features of individual circumstances.

(d) Did the model succeed in predicting lengthened IBI with increasing numbers of live children merely because the !Kung, like most human populations, become less fertile as they get older? Probably not, because short 'replacement' intervals occur at any time (Howell, 1979), and intervals after the fourth interval are not longer than previous intervals; if anything they are shorter.

(e) All hunter gatherers have long IBIs (or so it is said). So was such a detailed model, so specific to !Kung life, really appropriate? Some hunter gatherers have much shorter IBIs than the !Kung, so the premise of the question is wrong. Furthermore, changing the model, for instance by omitting weight of food, gave different and less successful predictions.

These last two points share the implication that such a specific and quantitative model was unnecessary, that many other models might have predicted the same outcomes, and that many other populations would have confirmed the predictions. These are important reasons for exploring the wider applicability of this sociobiological approach to modelling costs of children.

14.3 CURRENT RESEARCH ON ANOTHER POPULATION

The backload model was never intended to apply to anyone but the Dobe !Kung. What would happen if one began a similar study in another hunter-gatherer population? If circumstances differed would modelling mother's costs predict different schedules of reproduction? Or was there some factor general to all hunter-gatherer populations that accounts for long IBI and that was neglected in the backload model?

Blurton Jones *et al.*, (in prep.) report preliminary analyses of their field-work on Hadza women's work and child rearing, and compare their findings with the !Kung studies. Hadza women leave children in camp at a younger age than !Kung women do, and wean them soon after this (about 2.5 years old). Hadza children forage quite effectively. From about 5 years old they are capable of obtaining about half their food by their own efforts and probably do so much of the time. Women carrying babies may be less efficient at procuring food than women not carrying babies. Older women, past child bearing age, work hardest and bring in more food than anyone else (Hawkes *et al.*, in prep.).

The climate (less or equally severe, depending on which heat stress index is used), and shorter distances walked, suggest that backload could not be a serious limit for Hadza women. They often carry about 15 kg (food plus a baby) but only for 10–30 minutes. Thus the Hadza appear to lack some of the constraints acting on !Kung women. If we nonetheless considered backload as the most important cost for Hazda women then we can insert Hazda values in the Blurton Jones and Sibly backload model. The Hadza version of Figure 14.3 shows a gentler slope with its upturn at a shorter IBI, thus we would expect shorter IBI in the Hadza. Dyson's (1977) demographic analysis of the Hadza, using data from the International Biological Program expeditions in 1966 and 1967, suggests higher completed family size. Thus if reproductive careers are of comparable length in Hadza and !Kung (they may not be), the Hadza must have somewhat shorter IBIs than the !Kung.

However, Hadza women forage for much longer hours (more days of the week) than !Kung women and they may run short of time. The possible inefficiency of women with small babies suggests that a model which attends to the effect of IBI on mother's efficiency and thus the time taken to feed the resulting family would be the most appropriate model for Hadza women's child rearing costs. The model is not yet constructed but at this point it seems that the best model of costs for the Hadza would be composed of different costs than the successful !Kung model.

The Hadza research makes the !Kung seem even more interesting and Blurton Jones and Sibly's 'givens' even more challenging. Why do not !Kung children forage more? Why don't !Kung women leave children home more often, even if they had to leave them unattended like Hadza children? Hadza do all these 'commonsense' things, so why don't !Kung?

The low cost of Hadza children, and the help they sometimes give with feeding younger relatives suggests that beyond a certain age, they may actually make it easier for the mother to rear other children. This reminds one of the common argument in anthropology that children are an economic benefit in simple farming societies, and that this accounts for the high fertility in such families – because more children make a person wealthier, not because having more children is a goal in itself. It would be impossible to show an effect of Hadza children on family wealth for bush living Hadza accumulate no wealth. But the benefit of Hadza children could eventually be directly measured in time and calories, and translated into the reproductive terms of the effect this amount of time and energy has upon the mother's opportunity to produce more children. However, I now wish to discuss costs of children in a more simplified and generalized way.

14.4 GENERALIZING TO A WIDER VARIETY OF POPULATIONS

14.4.1 The shapes of curves of cost of children against age of child

Let us consider three theoretical curves (Figure 14.6) of costs of a child such as a cost of acquiring food for the child, not necessarily backload or time. (This discussion excludes once in a lifetime costs such as using up a lifetime supply of egg cells, or a fat store or non-renewable resource. The argument may be restricted to costs such as the cost of acquiring renewable resources

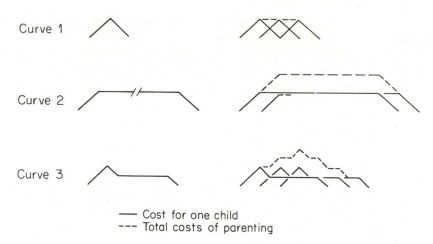

— Cost for one child
- - - Total costs of parenting

Figure 14.6 Some hypothetical curves of costs of children against age, and ways in which multiple births might accumulate these costs. Vertical axis represents cost of child. Horizontal axis represents time. On the left is the theoretical curve of cost of one child against its age. On the right is the sum of these costs for successive births, showing the level of cost sustained by the mother, with horizontal axis representing a segment of her reproductive career.

needed to raise the child, which can be acquired only at a limited rate.) Curve 1 climbs sharply and falls sharply after the birth of the child. This resembles a simplified form of the backload model, particularly the backload model with weight of food omitted. Curve 3 resembles the full backload model in which food is included. Curve 2 climbs to a low level at the beginning of the child's life and stays at this level for many years, falling only when the child is a young adult. We could imagine a fourth in which children are an economic benefit and the cost curve falls below zero for those times when the children are economically advantageous to the parents.

Let us explore how to schedule a family while varying these assumptions about costs of children but while keeping the same other assumptions as in the backload model: (1) there is an upper limit to parental cost which it is disadvantageous to exceed, (2) timing of births should be adjusted to pack as many children into the reproductive career as possible, as long as parental cost stays below the upper limit.

Under these assumptions the way to pack children whose costs follow Curves 1 and 3 is to have a continuing series of births, spaced so as to keep total cost just below the limit. The birth spacing will depend on the rate of fall of the curves. If substantial costs continue into middle childhood (Curve 3) spacing must be long and costs will climb as the reproductive career progresses. The curves determine rate of production of offspring. Total number of offspring will also depend on how long the reproductive process continues, i.e. on the number of births.

This serial strategy has no advantage in the case of Curve 2. The cost of these children does not fall. They could be raised successively but only by radically changing the dimensions of major life history stages, such as doubling the length of the reproductive career. Almost any IBI leaves the parent caring for several children simultaneously for a long period, with accordingly high costs. Birth spacing is almost irrelevant to maximizing reproductive success in such a case. Very short intervals are no worse than longer ones. The issue is simply one of how large the costs are and how high the limit to bearable costs is. Oddly enough the length of the reproductive career also becomes rather immaterial in this case, quite unlike the cases in which a series of births is the best strategy. Such a situation would lead to low fertility.

I know of no population for which such a situation has been described. One could speculate that child bearing in industrial societies (one to three quick births then no more) might be a response to Curve 2. In financial terms, children in industrial societies remain costly until adulthood. We have no good reason to expect the middle class of industrial societies to show adaptive strategies! But at least this framework makes the reproductive performance of the industrial middle class seem a little less bizarre, perhaps not altogether outside the range of possible adaptive strategies.

Consideration of these curves shows that several of the predictions that I derived from the backload model originated from the shape of the curves (optimal long IBI, lengthening IBI with more children, short first IBI independent of mortality). Others depend more on the nature of the costs – we expect women living at cattleposts to be different because they will not be gathering so much heavy food, so our expectation depends on what the cost is, not how it changes as the child grows.

If the Hadza time costs followed the same curve as the !Kung weight costs, then several of the predictions would be the same: e.g. longer IBI as the number of surviving children in a family increases. In predicting circumstances that change reproduction, however, we would look for influences on time, not on backload. A machine for grinding baobab might shorten Hadza IBI, whereas a donkey for carrying loads might shorten !Kung IBI (Lee (1979) reports the use of donkeys for this purpose by the !Kung).

14.4.2 Production versus survivorship: how often do we expect long IBIs to pay off better than short IBIs?

We could think of parents as dividing the resources available for reproduction between (1) production of offspring, (2) survivorship of offspring, and (3) enhancing the reproductive success of their offspring. Let us first consider the trade off between production and survivorship.

Commonsense suggests that reproductive success would normally be maximized by maximizing the rate of births. But if, as we showed for the !Kung, short intervals lead to such an increase in mortality that more descendants are lost than are gained, it will pay to reduce rate of births. Simple arithmetic can be used to explore the level and distribution of mortality that would select for reducing the rate of births (increasing IBI).

If we suppose a constant reproductive span, we can calculate the number of births that arise from any inter-birth interval. Thus if a woman kept to 20 month intervals throughout a career of 20 years (240 months) she would give birth 12 times after her first birth. If she maintained 60 month intervals she would give birth only 4 times after the first, for a total of five births.

Next let us calculate the combinations of IBI and mortality that give equal reproductive success. Suppose a woman is to end up with three surviving children. We can calculate the percentage of these children that she could lose while still ending up by producing 3 surviving offspring from each schedule or inter-birth intervals. The woman on the 60 month schedule can afford to lose only 2 children, a percentage loss of $2/5 = 40\%$, and will still raise 3 children. The woman on the 20 month schedule can afford to lose 10 children, a percentage loss of $10.13 = 77\%$ while still raising 3 children. I calculated mortalities in this way for each IBI schedule to achieve 2, 3, 4, 6, and 8 surviving children. Since the criterion age for survival was ten years old, and since the resulting lines represent equal 'yield' teenagers I called them 'isoteens' and some (labelled 2, 4, 8) are plotted on Figure 14.7.

The plot of isoteens illustrates some important points about optimal inter-birth intervals. If mortality depends on IBI, and mortality plotted against IBI follows a steep straight line (line A on Figure 14.7), then the optimal interval (the point where it is nearest to a higher isoteen) will be the one where the line reaches zero mortality. If mortality follows a flatter line (such as line B in Figure 14.7) then the shortest possible IBI yields most descendants, even though mortality will be higher. A concave curve of mortality against IBI will give an optimal IBI somewhere above zero mortality at the point where the curve reaches the highest isoteen. This is illustrated by Curve C in Figure 14.7.

For an IBI longer than the physiological minimum of 9 months to pay off, the curve of mortality against IBI must at some point be steep enough to cross an 'isoteen'. Examining the area of higher isoteens shows that, in a population with low mortality, only an extreme increase in mortality with shortening IBI can produce an advantage from longer IBI. Consequently, only very high mortalities and extremely steep curves of child mortality against IBI will render long intervals more productive than short intervals. We must then expect long intervals to be unusual over the range of human ecologies. It is possible that curves of mortality against IBI are seldom straight, and show a sharp increase at extremely short intervals under a wide range of circum-

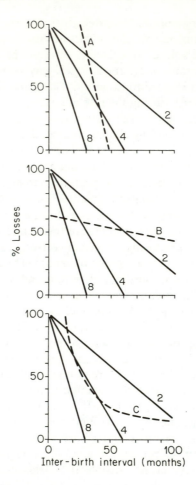

Figure 14.7 Graphs of 'isoteens', combinations of mortality and IBI that give the same number of surviving offspring.

stances. But it is clear that the !Kung findings should not be over-generalized.

It is thus not surprising that Hobcraft *et al.* (1983) comment in their study of birth spacing and mortality from the World Fertility Survey that 'For what it is worth, we note that any family trying to achieve maximal numbers of surviving children at any cost would, in the light of these results, continue to bear children at the most rapid rate possible. The dramatic excess mortality is not enough to negate the extra births. However, it is hard to recommend a pattern with such disastrous human consequences.'

The last sentence reminds us that an incidental conclusion from this arithmetic is that the overall level of mortality in a population may be selected to be high (as Daly and Wilson (1983) also point out). Differences in mortality may not directly reflect the ability to avoid mortality but will reflect the more indirect factors that determine the optimal trade-off between mortality and fecundity.

14.4.3 Enhancing reproductive success of offspring

Parents can promote the reproductive success of their children by providing adequate nutrition (more than is required for survivorship?), by developing the children's competitive and subsistence abilities, by endowing them with defendable wealth, by direct help with feeding and caring for the grandchildren, and by help in finding mates. It will obviously be worth doing this if little cost to production and survivorship of offspring is entailed. Will it ever be worth substantially reducing own production or survivorship in order to increase the reproductive success of a smaller number of offspring? For this to happen the benefit to the parent's fitness from the increase in offspring's reproductive success has to exceed the costs to parent's fitness of any sacrificed offspring.

The direct study of grandparents would be one context in which to study this issue. We know that grandparental care occurs. We do not know that the benefits of grandparental care outweigh the loss in production of further offspring. Another context in which to look for reduced fertility might be in populations where reproductive success is very much influenced by access to limited, defendable and inheritable resources such as land or livestock (as anthropologists have often remarked). Even under these circumstances, reduction in number of offspring might only occur, as illustrated below, when fitness returns from the resource increase with amount of resource.

We can illustrate this argument by imagining the following. Suppose 100 acres of farm will feed 8 children but 50 acres will only feed 3, and 25 acres only 1. A man has 100 acres, raises 4 children and divides the farm evenly between them. They get 25 acres each, thus each child can raise only one child on its 25 acres so our man will get only 4 grandchildren. If our man raises only 2 children and divides the farm between them they each get 50 acres. Each of these 2 children can raise 3 children on his 50 acres, so our man will have 6 grandchildren. If our man raises only one child this child can raise 8 children on his 100 acres and thus our man will get 8 grandchildren (the next generation would be much reduced however, and there is a question about how the interests of successive generations would be resolved). (More often we observe cases where the parent gives the entire farm to only one child, and lets his other children take their luck with alternative strategies such as emigrating, which may increase his possible number of grandchildren even more.) If the number of children fed per acre was constant then our man would get the same number of grandchildren however many children he divided the farm among: 4 children, 24 acres each, 2 children per 25 acres, 4 × 2 = 8 grandchildren; 2 children, 50 acres each, 4 children per 50 acres, 2 × 4 = 8 grandchildren. The message of this incomplete exercise is that it is only when the fitness returns of the resource increase with scale (when disproportionately more children can be raised on the bigger farm) that there will be a fitness payoff from limiting the number of children that the farm is divided among.

Herds of livestock whose crop may be expected to be very dependent on their size (given the 'geometric' rate of increase of animal populations), may be a more likely source of real cases of this type. The effect has to be large enough to outweigh the lower relatedness of grandchildren than children to this original parent.

14.5 DISCUSSION AND CONCLUSIONS

Though not without problems, looking at fitness maximization, costs of child rearing and scheduling of families seems to provoke many research questions, and to offer an understanding of some reproductive schedules far different from those of the much studied !Kung. It suggests three special circumstances in which we might expect reduced fertility:

14.5.1 Long IBI resulting from the relationship between mortality and IBI

The work on the !Kung was described and several criticisms of it were discussed briefly. Harpending and Pennington's (in prep.) recent analysis shows that a major component of !Kung reproductive success remains unexplained. Individuals that raise more children are individuals that give birth more times. This is also true of the Howell sample discussed in this paper, but does not negate the finding of optimum IBI.

If the relationship of mortality to IBI is weak, then it may pay the mother to produce many offspring, risking higher mortality. Conditions that favour the opposite, rearing few while keeping mortality low, will be uncommon. The !Kung situation, in which high mortality levels, together with the steep curve of mortality against IBI, determine that the strategy of long IBIs is more productive than a strategy of short IBIs, will be uncommon, Preliminary observations on the Hadza suggest that it may not even apply to other African savanna hunter-gatherers.

14.5.2 Costs of child that do not decline with age of child

Usually we have assumed that the costs of a child decline as it matures. In such a case, births would best be fitted into the reproductive career in succession, delaying the next birth until costs of raising the previous child had declined. If the costs of raising a child stay more or less constant as it grows up then there is no advantage in serial births and family size may be limited to the bearable costs, divided by the cost per child. Such a family might be very small compared to one in which the decline in child costs with age allowed sequential births, even though the parent is able to bear the same level of costs at any time in each case.

14.5.3 Investment in reproductive success of offspring

Another possible context for low productivity of offspring may be when parents are able to have a strong effect on the reproductive success of their offspring but are unable to divide his effect among many offspring. It seems probable that inherited wealth, and time or resource-consuming education might provide such a context. Direct grandparental care (provisioning or baby sitting) may be more effectively divided among greater numbers of sequential offspring. These issues must be explored further.

This sociobiological perspective on demography is of course not fully developed but there do seem to be some useful beginnings. This perspective includes fertility and mortality as integral parts of the reproductive strategy, where they are not treated as extraneous 'prime movers' as they tend to be in demography. The most obvious orientation of this approach is a reversal of our ethnocentric expectations. Ethnocentrically we cannot understand why people reproduce so fast as to degrade their environment, and why they refuse recommendations that would reduce child mortality to near zero. But as sociobiologists, we expect people to have as many children as they can, and we are not surprised by rapid population growth when wealth or medicine loosen traditional constraints. The conversion of resources into descendants can take many forms. Changes may have paradoxical effects – making it easier to raise a child may lead to shorter IBI and higher mortality and not to the lower mortality that the intervener hoped for. Western medicine aims at zero child mortality. The opposition to it may be something much more profound than any imagined ignorance, superstition or wickedness of parents.

ACKNOWLEDGEMENTS

I wish to thank Nancy Howell for access to her data and for advice on its nature, and Henry Harpending for discussion of his findings.

REFERENCES

Blurton Jones N.G. (1986) Bushman birth spacing: a test for optimal interbirth intervals. *Ethology and Sociobiology,* **8**, 135–42.

Blurton Jones N.G. (1987) Bushman birth spacing: direct tests of some simple predictions. *Ethology and Sociobiology,* **8**, 183–203.

Blurton Jones, N.G. and Sibly, R.M. (1978) Testing adaptiveness of culturally determined behavior: Do bushman women maximise their reproductive success by spacing births widely and foraging seldom? *Society for Study of Human Biology Symposium 18: Human Behavior and Adaptation.* Taylor and Francis, London.

Blurton Jones, N.G., Hawkes, K. and O'Connell, J.F. (in prep.). Measuring and modelling costs of children in two foraging societies: implications for schedule of reproduction (eds V. Standen and R. Foley). In: *Comparative Socioecology of Mammals and Humans.* Blackwell, Oxford.

Daly, M. and Wilson, M. (1983) *Sex, Evolution and Behavior.* Willard Grant Press, Boston.

Dixon, W.J. (ed) (1981) BMDP Statistical Software 1981. University of California Press, Berkeley, Calif.

Dobbing, J. (ed.) (1985) *Maternal Nutrition and Lactational Infertility.* Nestlé Nutrition Workshop Series. Volume 9. Raven Press, New York.

Draper, P. (1976) Social and economic constraints on child life among the !Kung. In: *Kalahari Hunter-gatherers* (eds, R.B. Lee, and DeVore, I.), Harvard University Press, Cambridge, Mass.

Dyson, T. (1977) *The Demography of the Hadza in Historical Perspective.* African Historical Demography, Centre for African Studies, University of Edinburgh.

Ellison, P.T., Peacock, N.R. and Lager, K. (1986) Salivary progesterone and luteal function in two low fertility populations of northeast Zaire. *Hum. Biol.,* **12**, 169–78.

Harpending, H. and Pennington, R. (in prep.) Fitness and Fertility among Kalahari !Kung. *Amer. J. Physical Anthropology* (in press).

Hawkes, K., O'Connel, J. and Blurton Jones, N.G. (in prep.) Hardworking Hadza grandmothers (eds V. Standen and R. Foley). In: *Comparative Socioecology of Mammals and Humans.* Blackwell, Oxford.

Hobcraft, J., McDonald, J.W. and Rutstein, S. (1983) Child-spacing effects on infant and early child mortality. *Population Index,* **49**, 584–618.

Howell, N. (1979) *Demography of the Dobe !Kung.* Academic Press, New York.

Howie, P.H. and McNeilly, A.S. (1982) Effects of breast feeding patterns on human birth intervals. *J. Reprod. Fertility,* **65**, 545–57.

Konner, M.J. and Worthman, C. (1981) Nursing frequency, gonadal function, and birth spacing among !Kung hunter-gatherers. *Science,* **207**, 788–91.

Lee, R.B. (1972) Population growth and the beginnings of sedentary life among the !Kung bushmen. In: *Population Growth: Anthropological Implications,* (ed. B. Spooner), M.I.T. Press, Cambridge, Mass.

Lee, R.B. (1979) *The !Kung San. Men, Women, and Work in a Foraging Society.* Cambridge University Press, Cambridge.

Nag, M., White, B.N.F. and Peet, R.C. (1978) An anthropological approach to the study of the economic value of children in Java and Nepal. *Current Anthropology,* **19**, 293–306.

Index

Index